Multi-application Smart Cards
Technology and Applications

Multi-application smart cards have yet to realise their enormous potential, partly because few people understand the technology, market and behavioural issues involved. Here, Mike Hendry sets out to fill this gap with a comprehensive guide to the technology, business and implementation aspects of this pivotal technology.

Following a review of the state of the art in smart card technology, the book describes the business requirements of each smart-card-using sector, and the applications and support systems required to sustain multiple applications. Implementation aspects, including security, are treated in detail and numerous international case studies cover identity, telecoms, banking and transportation applications. Lessons are drawn from these projects to help deliver more successful applications in the future.

Invaluable for users and those responsible for specifying, evaluating and integrating multi-application systems, this book will also be useful to terminal, card and system designers, network, IT and security managers and software specialists.

Mike Hendry is a freelance consultant and industry expert on cards and payment systems. He has many years of international experience in industry, was the Technical and Operations Director of the UK Chip and PIN Programme, and is also the author of several books.

Multi-application Smart Cards

Technology and Applications

Mike Hendry

CAMBRIDGE UNIVERSITY PRESS
Cambridge, New York, Melbourne, Madrid, Cape Town, Singapore, São Paulo

Cambridge University Press
The Edinburgh Building, Cambridge CB2 8RU, UK

Published in the United States of America by Cambridge University Press, New York

www.cambridge.org
Information on this title: www.cambridge.org/9780521873840

First published 2007

Printed in the United Kingdom at the University Press, Cambridge

A catalogue record for this publication is available from the British Library

ISBN 978-0-521-87384-0 hardback

Contents

Foreword

Smart cards are a thriving industry!

How do we know this to be the case? Well, if we look at the landscape of a typical industry sector, we see in smart cards the same characteristics we would witness in any other established and mature market.

For instance, companies have been created and thrive financially, based solely on the technology itself. These companies compete fiercely for a market share and brand leadership. Aggressive actions, such as mergers and acquisitions, and rigorous oversight of intellectual property rights are commonplace in the quest to increase both the industry and shareholder value. Dedicated industry analysts have built careers by following market movements and advances in the technology, and by prognosticating its future potential.

Trade shows and events have been established in every region of the world, dedicated to the exhibition of the technology and the sharing of information and industry best practices. These highly specialised gatherings not only showcase the latest in smart-card technology, but carefully articulate its relevance to critical sectors such as government, financial, retail, transit, healthcare and mobile telecommunications.

Industry associations have emerged to develop standards for smart cards and the applications that depend on the technology. In addition to developing standards, these birds-of-a-feather organisations have become valuable forums for information exchange between technology providers and end-user communities.

Magazines, periodicals, newsletters and websites cater exclusively to the smart-card industry. At the time of writing, a Google search on 'smart cards' resulted in 92 500 000 possible sites to explore.

Clearly, there is much to communicate.

So, by all accounts, one would assume that, in the presence of this much information and activity, the smart-card industry is, indeed, established and mature. While this may be true of some market applications (GSM and EMV) and within some regions of the world (Europe and Asia), smart-card technology has not been woven into the fabric of every card-holder's daily life on a global basis – at least not yet.

Ultimately, however, the widespread and pervasive use of smart-card technology is not only predictable, but inevitable.

Driven by the need for security, smart-card technology *will* emerge as the platform of choice in key vertical markets. Driven by the need for card-holder acquisition and retention, smart-card technology *will* emerge as the platform of choice to deliver enriched applications and services aligned with individual card-holders' lifestyles. And, driven by the need to remain competitive, smart-card technology *will* continue to drop in price as standards evolve and commoditisation of the technology continues.

With these drivers in mind, the smart-card industry is taking action to capitalise on market momentum and the unique value proposition of this singular technology. Without question, the smart-card industry is on a trend to move:

From several operating system models to a dominant few. This will drive economies of scale in the industry and drive down cost without hindering competition.

From single purpose, single application cards to multi-application cards. This will enhance the card-holder and issuer value proposition for smart cards and expand the business model, justifying the investment in the technology.

From contact technology to contactless and multiple interface (TCP/IP, NFC, etc.) technology. This will enhance the utility of a single card platform by allowing its integration into different use environments.

From a card form to alternative forms such as smart objects and tokens. This will allow differentiated marketing practices among issuers and broaden the commercial appeal of the technology in several consumer-based application areas.

From single issuer models to co-operative private-public sector partnerships. The flexible nature of the technology will encourage novel commercial arrangements to emerge.

And from confusion to clarity on how to achieve interoperability. This will expand the size of the overall marketplace and provide a sustainable environment necessary for the rapid and widespread proliferation of the technology.

As these trends unfold over time, the utility of the smart card will multiply and play a significant rôle in enhancing daily lives, while safeguarding the applications and personal information of individual card-holders.

In this book, Mike Hendry takes care to address not only smart-card technology, but the transformation of the industry as described above. His work adds to the industry's rich body of knowledge and provides a fair and balanced view of smart cards as they exist today.

This industry, however, like many industries, is in a constant state of change and evolution. As such, any book on smart cards (or any non-fiction book for that matter) will become historical in context. So, I encourage you to study the information in this book as a 'first step' in your journey to understand the technology and the industry.

The next step – and on an on-going basis – will be to immerse yourself in the ebb and flow of the industry itself, through participation in industry associations, attendance at trade shows and events and by reading the periodicals and future new releases of books that cater to this subject.

Multi-application Smart Cards: Technology and Applications offers a foundation upon which to build your on-going involvement in the smart-card industry.

Welcome to our industry and to our unique technology!

Kevin Gillick
Executive Director, GlobalPlatform

Acknowledgements

Writing a book about multi-application smart cards is a microcosm of a multi-application smart-card project – although it demands of the author an understanding of the key issues, particularly those that affect structure and organisation, successful delivery requires input from a wide range of experts in different technology and application fields.

I have been fortunate to be able to draw on the experience of many such experts from different industries, viewpoints and countries. I greatly appreciated the early support, guidance and encouragement of Richard Poynder at the UK Smart Card Club, Greg Pote of the Asia-Pacific Smart Card Association and Ayse Korgav, Tono Aspinall and Kevin Gillick at GlobalPlatform; I am delighted that Kevin has provided a foreword.

Several extremely busy project managers and project owners have taken the time to contribute case studies or material for case studies, or to review case studies I have written; thanks are due to Erik Wellen and Dr Maan Kousa at King Fahd University for Petroleum and Minerals, Jusuk Lee of IBM Korea, Jason Lee at SK Telecom and Professor Ho Geun Lee of Yonsei School of Business, Dave Taylor at Barclaycard Business, James Lu in Taiwan, Wong Wan Ling at Welcome Real-Time, João Miguel Almeida of Link Consulting, Chua Siew Ling of QB, Chang Yun Chang at FEETC, Elvin Huang and Mike Cowen of MasterCard, Raymond Wong and T. K. Wong at the Hong Kong Immigration Department, Martin Arndt of the Royal Oman Police, Richard Pinnick at Fortress GB, Richard Poynder again, and Eddy Cheah and Wan Mohammad Ariffin in Malaysia.

For chapter reviews and advice in specific domains I have to thank Ian Duthie, Chris Shire, Julian Ashbourn, Bill Reding, Marc Kekicheff, Tim France-Massey, Ian Volans, Peter Jones and Peter Stoddart.

It has been a pleasure to work again with Julie Lancashire at Cambridge University Press, where the production team has also been very helpful in ensuring the quality of the final product. Meanwhile James Lu and Dr Lyndon Huang are working hard on the Chinese translation of the book, which will be published by the Taiwan Academy of Banking and Finance, where I also thank Emily Kuo and Erica Lin.

Lastly, but perhaps most of all, I would like to acknowledge the input of the hundreds of people all over the world with whom I have worked on smart-card projects and whose views and insights have helped to form my own.

My wife Valerie has tolerated the interruption to my work, sleeping patterns and mealtimes that any international book production involves; I never cease to be grateful for her understanding and support.

Shepperton, July 2006

Part I

Introduction

1 Background

1.1 Smart cards in daily life

Cards are so much part of our daily lives that we do not even think about their functions, the technology behind them or the things that make them special.

Cards are behind some of the biggest changes in behaviour in the Western world since 1970 – the way we enter buildings, pay for goods in shops, speak to our friends and business partners. Back in 1970 it would have been difficult to imagine the ease with which we now draw money from ATMs in foreign countries, or that many ten-year-olds would have their own telephones.

Many of these changes have helped to spread technology as well, benefiting a wide range of people in poorer countries and remote areas. Where there is no reliable telecommunications network, the ability to store a patient's health records in a card can save lives. In most African and many Asian countries, there are many more mobile telephones than fixed lines; these telephones not only use cards to provide security and added functions, but may themselves act as terminals for other card-based applications, such as microfinance.

It's not all good news, of course: some of these changes have been made necessary by the increasing need for security, while others have increased efficiency but have incurred a cost in reduced personal service and social interaction. And not all card projects have been equally successful.

Most users do not need to think about how they work, in much the same way as an artist or writer does not need to think about the pencil he or she uses. Occasionally, though, it is worth thinking carefully about cards: how we use them, their functions and technology, and how we can use them to improve life and business. Not only can those involved in the cards business benefit from this reflection, but also everyone who manages, controls, designs or operates a business in which cards are used – and that means virtually every business.

This book aims to provide some of the background as to how businesses can use cards more effectively. In particular, it focuses on the most advanced card technology in use today: the multi-application smart card. But along the way we will see some situations in which much simpler card technology may do just as good a job.

1.2   Card functions

1.2.1   From identification...

When people do think about cards, it is mostly about the cards in their purse or wallet: their bank cards, office access cards, health cards or identity cards. Nearly all of these cards have *identity* as their main function – they help us to identify ourselves to a system.

Where we are identifying ourselves to a person, the simplest solution is a 'flash card' (carrying a photograph or just a name and signature); these identities are rarely checked for validity. Where we are identifying ourselves to a machine, or where there will be a check against a database, the most widely used technologies are magnetic stripes and bar codes, both of which work extremely well. We can also use contactless (wireless) technologies such as Wiegand tags (the rather thicker cards often used for office access) or contactless smart cards.

Objects can benefit from identification too – goods in industrial processes and in the retail supply chain have traditionally been identified using bar codes, while there is a growing use of radio-frequency identification (RFID) tags (see Figure 1.1) to track high-value goods and prevent theft in shops, and in libraries to track books.

These tags only transmit a string of data to the reader. In fact, the characteristic of all the technologies described so far is that they are read-only: they provide a reference number that allows a system to access a record in a database.

1.2.2   ...to authentication

Increasingly, though, the real requirement is not only to *identify* the record in the database, but also to *prove* that the person presenting the card is actually the person referenced by the database, or that the card itself has not been forged or altered. This is where the smart card starts to come into its own.

Smart cards not only contain data but can also perform operations, such as comparing the data with an external source, computing an electronic signature from some data or incrementing and decrementing counters. Their other important feature is their ability to store data in secret areas that cannot be accessed from outside the card, but only by the software on the card – this makes them particularly good for cryptographic purposes such as protecting the confidentiality of encrypted files, or providing proof of a transaction.

Where a requirement for identification includes an underlying requirement for authentication, smart cards are usually the best tool for the job. So smart cards are now the standard form of bank card in Europe and Asia, and increasingly in other parts of the world, to prove that the card has not been copied or altered, and that the user is the correct card-holder. There is a smart card in every GSM mobile telephone, to authenticate the account and protect the confidentiality of the conversation.

Where governments issue machine-readable passports or identity cards, the only generally accepted way to prove reliably that the card and the data on it are genuine is to use a smart card: these are now the standard form of identity card for 7 million

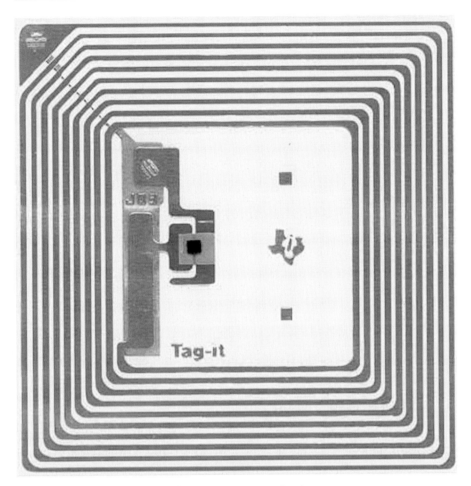

Figure 1.1 RFID tag (courtesy of Texas Instruments, Inc.)

Hong Kong residents, 23 million Malaysians and even 4 million US service personnel and Department of Defense employees. Over the next ten years, most countries will adopt this technology for inclusion in passports.

1.2.3 Data storage

Although it may seem obvious that we use smart cards to store data, actually this is not usually a core function of a card. Most smart cards have a quite small data storage capacity (a few kilobytes, or tens of kilobytes at the most), and when in use they are always connected to a terminal or reader, which almost certainly has a much greater storage capacity and in turn is connected to a server with access to a database.

So smart cards are used when the storage performs a special function: for example, when the data belong to the card-holder and not to the system operator the card can protect the card-holder's privacy and ownership rights. In a GSM telephone, the SIM card stores data and applications that are under the control of the network operator, not the handset manufacturer. With a biometric application, storing and comparing the template on the card ensures that it cannot be copied or inspected whilst being transmitted from the host system to the terminal.

1.3   Advanced applications

1.3.1   Cryptography

Although the notion of keys and secrets underpins almost every smart-card application – even the most basic memory card today uses keys to control reading and writing of the data on the card – relatively few cards actually have the special hardware needed for fast computation of the variable-length arithmetic functions and public-key algorithms used in modern cryptography.

Nonetheless, smart cards play an important rôle in many cryptographic systems; they provide the token that represents the physical factor in multi-factor authentication (these terms and concepts will be covered in Chapter 4). It is equally important to recognise that the card is only a part of such a system and cannot provide security on its own; many critical analyses of smart-card security have missed this point.

Chapter 5 will address the ways that smart cards can help to provide the five main security services, and the standards that smart-card-based security should follow.

1.3.2   Database access and linking

Although smart cards themselves are an expensive and limited storage medium by modern standards, we are moving towards a world in which almost every piece of information is available in some form of online database; the problem is to identify the data we are seeking and to control access to that information.

Smart cards can greatly assist in both of these functions: not only by carrying reference fields that point to a record in the database, but also by identifying the category to which the user belongs and their rights to view or change the data. This is very important, for example, in health records systems, where different groups of health professionals have different needs and rights to view patient records, or with government data, where data protection and separation of functions must be enforced.

1.3.3   Biometrics

Every form of identification has its drawbacks: cards can be stolen or shared, passwords and PINs can be viewed, signatures and photographs are very difficult to verify. But there are many situations in which it is very valuable to be able to verify (with a reasonably high degree of certainty) that people are who they claim to be.

A smart card is an ideal tool for this: a person can carry a card that records his or her identity, a biometric and a certificate linking the two. We can regard these as three statements:

- 'I am John Smith; my reference is 123456789.'
- 'This is my fingerprint (or iris scan, signature, etc.)'
- 'We (The US Government, Acme Manufacturing Corporation ...) assert that the person whose fingerprint matches this one is indeed John Smith, reference 123456789.'

Provided the person relying on this card trusts the US Government and knows its public key, this is a very dependable form of authentication.

We will see in Chapter 4 that many forms of biometric data can be stored very efficiently in a smart card.

1.3.4 Multiple applications

Smart cards are particularly valuable when they combine a number of functions. As we will see in the next chapter, there are several ways to create a multi-function card, but the smart card's unique ability to carry several different programs (normally known as card applications) makes them capable of performing a much wider range of functions than other card types.

For example, the Malaysian identity card (described in more detail in Chapter 17) also carries a driving licence, health services entitlement, electronic cash and a digital signature to permit signing of e-commerce transactions. Some Mexican bank customers can store the URLs of their favourite websites, together with their usernames and passwords, on their bank cards.

1.3.5 The universal helper

This general ability to carry out seemingly unrelated tasks has led to many futuristic descriptions of the single card that accompanies its holder through daily life, giving access to trains, buildings and computers, keeping appointments, making payments, and even making coffee to the user's exact taste.

The user, of course, would own such a card and determine what applications are loaded. It may not actually be card-shaped; it could be contained in a mobile phone, wristwatch, key-fob or ring.

Although these are all technically possible, this is not how users see their cards today. In practice, most cards embody a *relationship* of the holder with the card issuer: an employer, service company, bank or government. As we will see, there are several categories of use type and relationship, and the second half of this book is concerned with the ways in which those translate into business requirements for a smart-card project.

1.4 The smart-card business

After a slow start, the smart-card industry has grown steadily over the last 15 years as new applications have been developed using this technology and subsequently rolled out to new countries.

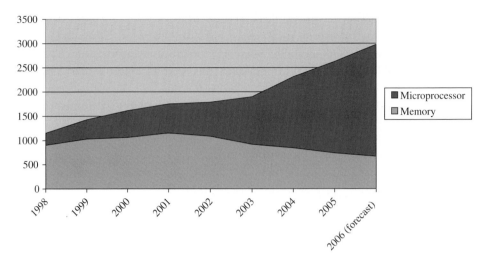

Figure 1.2 Growth in smart-card shipments 1998–2006 (source: Eurosmart)

Figure 1.2 shows the growth of smart-card shipments over the last eight years; this also shows how memory cards have become less important and microprocessor-based cards form a larger proportion of the total. The differences in the technology will be discussed in more detail in Chapter 3, but it is important to note that memory cards can still offer a level of security that is often adequate for closed systems: many transportation schemes, for example, will continue to use memory cards for years to come.

Figure 1.3 Smart-card market by region 1998 and 2005 (source: Eurosmart)

Figure 1.3 shows the way in which the geographic balance of the market has changed; whereas this technology originated in Europe, the main feature of the first few years of the 21st century has been the growth of Asian, and in particular Chinese-speaking, markets. China now accounts for one third of all SIM cards sold in the world, and the new generation of Chinese national ID cards will represent a further massive increase in this market over the next few years.

These developments have driven business at both the top and bottom ends of the market: in China, many new manufacturers have been established and are supplying a large proportion of local demand, for low-capacity SIM cards in particular. Countries like Taiwan, Korea and Malaysia, on the other hand, are the main motors for large multi-application cards.

Companies in the smart-card business have been through some rough periods: when demand was high, semiconductor supplies limited sales, and now as both demand and supply have grown, it has become difficult to maintain sufficient margins to pay for development of the new products that users are seeking.

The multi-application smart card, in particular, offers the opportunity for vendors throughout the supply chain to increase the value they add by creating more revenue streams, and at the same time to differentiate themselves from their competitors through both the products and services they offer.

1.5   Structure of this book

Part I sets the scene for the rest of the book: in Chapter 2 we look at the different ways of defining a multi-application smart-card scheme, and the different technologies that can be used to construct one. Chapter 3 covers the basics of smart cards and could be skipped by those who are already familiar with the technology.

Part II describes the different technologies used in multi-application smart cards, starting with two disciplines that are critical to most multi-application usage: bio-metrics and cryptography. In Chapter 6 we turn back to card technology, and look in more detail at the current state of the art, particularly in relation to microcontrollers and interfaces, including contactless interfaces. Chapter 7 does the same thing for reader and terminal technology, then Chapter 8 looks in more detail at the multi-application functions built into the core ISO 7816 family. Chapters 9 and 10 describe the two main families of multi-application operating system: JavaCard – GlobalPlatform and Multos, and in Chapter 11 we look at other operating systems with specific multi-application features. Chapter 12 addresses the complexities of card management systems for multi-application cards, and in particular what functions are needed in a card management system according to the type of scheme.

Part III is concerned with applications. It covers, in turn, the main application sectors for smart cards: telecommunications, banking and finance, transportation and government, and closes with those applications used by corporations, as well as universities, schools and other organisations for 'campus cards' and other closed user-group schemes. This part of the book looks at a very wide range of applications, including many that are today most often implemented on single-application cards. It seeks to show the business drivers and relationships behind each application and the implementation barriers, so that users from other sectors can see whether these applications would be suitable for inclusion in their cards, and how these drivers and relationships go together to construct a business case and a specification for a multi-application card project.

The book closes with three chapters on implementing multi-application cards: Chapter 19 looks at how to organise and structure a project team, and the difficulties of working with unmatched structures. Chapter 20 draws some lessons on specification and project management, while Chapter 21 attempts to draw some conclusions as to the applications and sectors most likely to succeed in the coming years.

Throughout the book, there are many case studies drawn from practice, in most cases contributed by the project managers responsible. They cover a wide range of sectors, countries and technology solutions, but each has some useful lessons for would-be implementors and will, the author hopes, increase the chances of success for those willing to learn from history and experience.

The book also includes as Appendices a glossary, references for further reading and a list of standards relevant to multi-application smart cards.

2 When is a card multi-application?

The term 'multi-application card' is used in different ways by different groups of people: the marketing department sees the card in terms of selling features, the IT department according to the technologies used by the card, and the operations department looks at the number of processes the card supports.

This chapter explores the definitions of the term and sets the framework within which the remainder of the book will use it.

2.1 Single-function cards

Most smart cards have a single function. There is a simple reason for this: the card issuer has issued the card to solve a specific problem or to provide a specific service. The relationship between the card issuer and the card-holder is generally not complex, while most card issuers are in one well-defined business. So there is no reason for the card issuer to provide multiple functions on the card, which in most cases would add to the cost.

Smart cards are, in many cases, replacing a magnetic stripe or visual identification card, which generally had only one function. So, for example, the earliest smart cards were used mainly for public telephones: they held value that could be loaded by the telephone company and decremented by the user making calls, and this was their only function. Many schools issue cards to their pupils for recording attendance at classes, while companies issue cards to their employees for access to buildings. These cards need no further functions.

Most single-function cards are either memory cards or microprocessor cards with only a very small fixed program. To be very precise, we do need to be careful when we talk about cards where only one function is used, because in some cases the card itself is capable of performing other functions ... but that will become clear later.

2.2 Multi-function cards

From the card-holder's perspective, the card becomes more useful when it can perform more *functions*. This has nothing to do with the number of applications, since multiple functions can be provided in several other ways:

2.2.1   Card-based functions (wired-logic cards)

Many memory cards have been designed to perform a specific range of functions relevant to a particular application. For example, Sony's FeliCa™ range provides a set of functions relevant not only to transport ticketing, but also covering identification and stored value.

Wired-logic cards can be very versatile: for example, quite a simple set of functions will allow a single card to be used for storing and decrementing value (perhaps for use in vending machines, a canteen and photocopiers), for controlling access to buildings and computer systems and for making and storing bookings for a sports facility.

But wired-logic cards always depend on the terminal application to provide much of the functionality, and also the security of the scheme. The card's functions are, by definition, fixed and hence security depends on the keys, which must be shared between the card and the terminal or host system. So these cards are generally more appropriate for closed-circuit schemes, in which the card issuer also controls all the terminals and can manage all the security keys in the scheme directly.

2.2.2   Server-based functions

It is possible to provide many functions using a single card where the majority of the work (in particular any decisions) is performed by software on a host system. The card in this case need only provide a pointer to a record on the host, and can be very simple. Many campus cards work in this way.

This approach imposes two main limitations: the need for always-connected networks and authentication. The transaction performance will always be limited if the terminal must connect for every transaction, and in particular if the transaction has to include a dialogue between host and card (we will see in Chapter 3 that data transfer speeds are one of the limiting factors in card performance). And if the host or network is not available, then the transaction cannot proceed.

If the card does not play an active part in the process of identifying and authenticating users, it will always be prone to 'cloning' attacks, in which a near-exact copy of the card is made, possibly with a different photograph or signature on it. For a secure scheme, either the card must play an active rôle or the authentication dialogue between user and host becomes very lengthy.

This model is, therefore, not used on its own for systems requiring high availability, high performance or even moderately high security. But it can be used to add non-critical functions to an existing scheme that already offers adequate authentication; an example of this would be a loyalty scheme offered by a bank on top of an existing payment card scheme.

2.2.3   Multiple datasets

In some cases, an appearance of multiple functions may be given by a single card application, by using multiple datasets. Bank payment cards following the EMV standard (which will be discussed in more detail in Chapter 15) use a common program capable of accessing several different files; there is a secondary application selection process that allows the customer or retail terminal to select between a debit or credit account, or to offer a wider range of functions in the card-holder's own country than abroad. Parameters in the different datasets simply enable different functions within the EMV application.

2.3   Multiple applications

From the perspective of the card programmer or developer, a multi-application card is one with several applications (programs) loaded in the card's memory. Again, there are several variants on this theme:

2.3.1   Distinct and co-operative applications

Most smart-card applications hold their own data, and when there are two applications on the same card, they do not share data – the terminal simply selects one or the other. It is, though, possible to arrange for applications to share data through the file structure on the card; most card operating systems have specific provisions for allowing both shared and private data.

As we will see in a moment, sharing data is really the only way that applications can co-operate.

2.3.2   Application selection

The ISO 7816 part 4 standard followed by most smart cards defines a way to select applications, starting from the answer-to-reset (ATR) signal that is sent by the card as part of the power-up sequence. In the ATR, the card declares the speeds and protocols it follows, and the terminal uses its response to change to a higher speed or protocol if it can support them.

The terminal then selects an application, either directly or by asking the card for a list of applications. It will usually know from the context of the transaction what action it wants to perform, and can select the appropriate application. In some cases, though, it may need to discover more details about the card before selecting an application; for example, during a retail transaction it may be necessary to select a loyalty application, if one is present, before making the payment. In this context, a multi-application card is one that announces several different applications.

Some card operating systems offer special functions for application selection – we will come to them shortly.

2.3.3   Application ownership

Where all the applications loaded on a card have the same owner, they can trust each other, and can usually be tested together. In this case, the management of the multiple applications is relatively simple and the need for special functions (for example, to manage security) is greatly reduced.

2.4   Operating systems

Where, on the other hand, there are several application owners sharing one card, or where the applications may not all be loaded at the start of the card's life, we are likely to need some kind of multi-application operating system, rather than or in addition to the proprietary applications (known as 'native operating systems') that most smart-card suppliers use to provide the ISO 7816-4 interface on their cards. Many people would consider that only cards using a multi-application operating system are 'true' multi-application cards.

Such an operating system has to perform several functions:

2.4.1   Application protection

A key task is to protect and firewall applications from one another, so that an error in one cannot affect any other, and one application can only interact with another in a strictly controlled and defined way.

2.4.2   Memory management

The operating system manages memory, allocating program memory (ROM and E^2PROM), data memory (usually E^2PROM) and working storage (RAM) to applications and ensuring that an application cannot access memory locations outside its allocated space.

It also controls the memory used for program storage, and ensures that applications can only be loaded into the memory space allocated to them.

2.4.3   Application downloading and updating

One of the most powerful features of a multi-application card is the ability to download or update applications after the card has been issued. Again, this is not a clear-cut distinction as many native card operating systems can accept patches and updates to programs, or the applications themselves can use a 'scripting' protocol that allows parameters and data to be updated during a normal transaction.

A multi-application operating system, though, gives much tighter control over the applications that can be downloaded, requiring or allowing applications to be downloaded in an encrypted form, and only decrypting them or allowing them to be run on receipt of a special command signed by the original card issuer.

2.4.4   Interpreter languages

Another potential benefit of using multi-application cards is the ability to develop applications in a language that is independent of the card type being used. Applications can then be developed once and run on many card types.

As we will see in Chapters 9 and 10, both the main current multi-application operating systems use a form of interpreted language to yield this benefit: Java in the case of GlobalPlatform and MEL for Multos.

2.4.5   'Open' features

Taking this argument further, full card and vendor independence requires the use of an open standard, in which the specifications are published and open to all. The current operating systems vary greatly in the extent to which they fit this definition.

2.5   Multiple organisations

Using a multi-application operating system opens up the possibility of sharing a card between organisations.

Even within one company or government department, it may be important to separate the data for different applications, and even to protect the applications from one another. For example, data protection and a need-to-know requirement may restrict certain data to the Human Resources department, or prevent a routine police patrol from accessing spent convictions. The critical corporate IT system access function must not be at risk of being overwritten by a data overflow on the new staff-satisfaction monitoring system.

Co-branding partners (such as a bank and a transport company, or a retailer and a telco) have even stronger reasons to protect their data and applications, and multi-application operating systems give them the tools to do this (or more specifically give the card issuer the tools to do this). Application owners can also use these tools to protect themselves from third-party suppliers, including application developers.

Although the technical tools exist to allow such sharing and protection, one of the themes of this book is that the technology alone is not enough: parties who want to work this way must align their objectives, priorities and timescales, their customer relationships and often even their cryptographic key structures in order to make this co-operation a success. This is a challenge for any two organisations; when those organisations are from different sectors, have different value-sets and are subject to different standards and regulatory controls it may be an insurmountable obstacle.

It is one of the reasons for the slow uptake of multi-application card systems; this book seeks to show what factors contribute to this difficulty and what steps can be taken to overcome it.

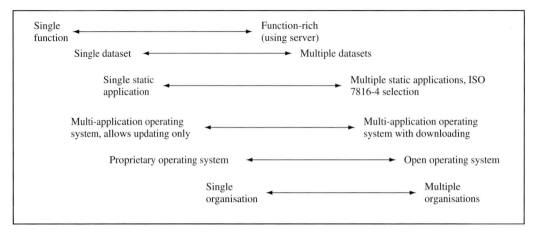

Figure 2.1 Definitions of a multi-application card

2.6   Conclusion

There is no single definition of a multi-application card; the distinctions are progressive, as shown in Figure 2.1, and the definition of a 'multi-application' card, therefore, depends on the context.

3   Smart-card basics

This chapter covers the key features of a smart card, its manufacturing process and the components of a smart-card system. It can be skipped by those who are already familiar with the technology and whose main interest is in advanced card types, and in particular in combining applications within a single card.

Appendix B also lists some further reading on smart-card technology in general.

3.1   What is a smart card?

3.1.1   Common features

A smart card is a card incorporating one or more integrated circuits within its thickness (see Figure 3.1). Smart cards are also often called chip cards or integrated circuit (IC) cards – these terms are interchangeable.

As we will see, the terms cover many cards that are not really 'smart' in the sense of being programmable, but the smartness comes from the way they are used as a part of a system.

Most smart cards meet the ISO 7810 standard (bank card size and thickness), but there are other standard card shapes, such as the ID-000 shape used by mobile telephone SIM cards. And some devices known as smart cards are not card-shaped at all – although this does raise a number of issues, as we will see in Chapter 6.

There are two main categories of smart cards, usually characterised as *memory* and *microprocessor* (or *microcontroller*) cards. The name microcontroller is technically more accurate since the chip includes memory, the serial interface and, possibly, more than one processor.

3.1.2   Memory and wired-logic cards

The simplest chip cards are not really 'smart' – they represent a portable storage device. Memory cards are viewed by the external application as data storage, and have a fixed and limited range of functions. Confusingly, in some cases these functions are actually provided by a microprocessor, but this is invisible to the outside world, since there is no way to vary the firmware and, hence, the functions of the card.

Today very few cards offer unprotected memory access, but these cards do exist and may be adequate for some small schemes where the security of the scheme does not

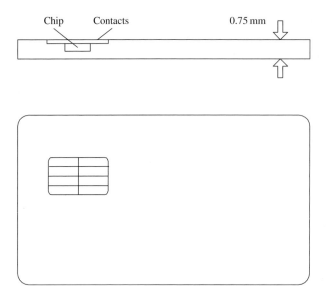

Figure 3.1 Smart card (or chip card)

depend on the card, which offers less security than a magnetic stripe card, but potentially more storage.

Much more common are wired-logic cards, in which access to the memory on the card is controlled by a security protocol, either involving encryption of the stored data or a password mechanism.

Such cards can be quite powerful and can allow multiple functions. For example, the NXP[1] MiFare™ and Sony FeliCa™ cards both divide their memory into sectors and fields; each sector has separate keys or access permissions for reading, writing, incrementing or decrementing the data in the field. Different readers, possibly owned by several organisations, may, therefore, have different permissions.

3.1.3   Microprocessor cards

A growing proportion of all smart cards are microprocessor cards. These benefit not only from added security, but also from a common interface based on ISO standards for both the transport and data layers (whereas wired-logic cards more often use proprietary protocols). Microcontroller-based cards have a definable software component (known as an application) and follow an application selection process that I will describe in Chapter 8.

For many applications, 8 bit cards (usually based on a variant of the Motorola 6805 architecture) are adequate: often only a few hundreds of bytes of data must be stored and transmitted, and there is no requirement for complex arithmetic or on-card cryptography. There are now several 16 bit and 32 bit cards available, and these are often preferred for

[1] Formerly Philips Semiconductors.

running multi-application operating systems. For sheer performance, however, it is often more appropriate to use a reduced-instruction-set (RISC) design or a separate crypto-coprocessor to handle the specialised arithmetic and shifting operations associated with the main cryptographic algorithms. These RISC processors have been optimised for the functions they will need to perform in a smart-card environment, while crypto-processors specialise in the variable-modulus arithmetic required by many public-key algorithms and the logical operations demanded by DES and similar symmetric algorithms.

Another way to improve performance in a smart card is to have separate processors that allow communications and memory access operations to take place in parallel with processing; the external communications for contact cards take place at relatively low speeds (most often 9600 bps, and a maximum of 38.4 kbps), and this is often the limiting factor in transaction performance.

For example, an application that uses 4 kB of data requires a minimum of 3.4 seconds just to transfer those data bytes to the reader at 9600 bps. The file structure will add some time – if the data are divided into many small files, then we must add the time required to select each file. And if the smart-card microcontroller does not have a separate communications processor then data transmission must stop while the device performs any file selection or processing operation. As a result, many smart-card operations require several seconds' transaction time.

3.1.4   Memory types and sizes

A microprocessor card normally has several types of memory:
- Read-only memory (ROM) for fixed programs and static chip personalisation data;
- Random access memory (RAM) for working storage;
- Electrically erasable programmable read-only memory (E^2PROM) for storing most data, program updates and some programs.

The E^2PROM usually accounts for most of the memory, and, in particular, the largest area of the chip and much of its power consumption. Typical memory sizes for modern microprocessor cards range from 4 to 64 kbytes.

An increasing number of products now make use of flash memory, which is non-volatile but rewriteable by pages, and takes up much less space and power than E^2PROM, raising the maximum memory available in a 25 mm^2 chip from 16–32 kbytes to several hundreds of kbytes (at 2006 levels).

Most smart-card chip manufacturers limit the size of their chips to 25 mm^2 because chips larger than this can more easily be damaged when the card is bent; this also limits the amount of power that can be used by the chip and that must be dissipated as heat.

3.2   Interfaces

Most smart cards – particularly in mobile telephony and banking – have a contact interface as shown in Figures 3.2 and 3.3. The connection to the chip is through a set of contacts on the front of the card.

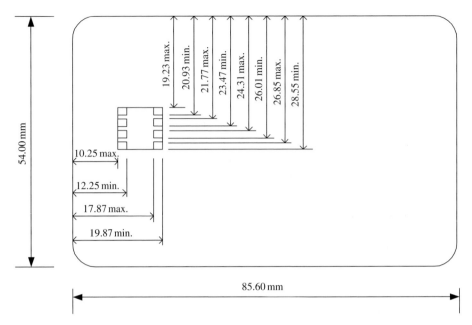

Figure 3.2 Contact positions and dimensions for ISO 7816-2 card

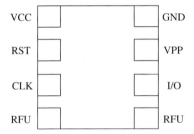

VCC			GND
RST			VPP
CLK			I/O
RFU			RFU

Figure 3.3 Contact assignments for ISO 7816-2 card

Cards used in transportation and access control are much more likely to use a contactless interface, where the energy to power the chip, and the communication between the chip and reader, is provided by radio waves from an antenna connected to the reader. This technology is used by an increasing number of cards, because of the speed and reliability it brings.

A third option, likely to be very attractive for multi-application cards, is to have both contact and contactless interfaces on the same card, with one or two chips.

3.2.1   Contact cards

The contact interface supplies power, clock and reset signals (known as VCC, CLK and RST), and also a single bi-directional serial interface (I/O). There is also provision for a sixth signal (VPP) to power the programming of the E^{2}PROM, but this is now rarely used.

The card and terminal follow a protocol to agree ('negotiate') a voltage and clock speed (the clock speed may determine the internal processing speed as well as the rate of data exchange between card and terminal). Most stand-alone cards work at the lowest specification: 5 V power and 9600 bps communications; however embedded cards (for example, in mobile phones) are much more likely to use lower voltages and higher communication speeds, where the terminal will support them.

Although, in principle, this negotiation is mutual – the card and terminal will use the best specification that they both meet – there are some implications here for multi-application cards in that cards designed primarily for low-voltage operation may be unreliable or have a shorter life at a higher voltage.

3.2.2   Contactless cards

Contactless cards use a radio-frequency (RF) field both to power the card and for communication between card and reader. The card incorporates an antenna (a few coils of wire wound buried in the card close to its outside edges) instead of the contact plate on the face of the card.

They are very widely used in public transport, and in fact have advantages over contact cards in many situations. For example, the absence of contacts means that cleanliness is not an issue. The physical action of inserting a card into a reader stresses the card and if done clumsily can damage it; these two factors mean that contactless cards are generally slightly more reliable than contact cards.

Interoperability for contactless cards is also more easily achieved. Because the card need only be located somewhere in the field where it can couple sufficient power, its shape and size are relatively unimportant, and the card will only generate sufficient power for its needs (within limits – an exceptionally strong field could cause it to over-supply itself).

3.2.3   Dual interface

For some applications, it is preferable to have both contact and contactless interfaces on the card. For example, many organisations prefer to use a contact interface for loading value onto a transport card, but a contactless interface for using that value. The international card schemes currently insist on a contact interface for all higher-value transactions.

For a passport or ID card, it may be felt better to load the data using a secure contact interface, even if the card when in use is always contactless. Many user organisations feel that contact interfaces are more secure than contactless interfaces, but in fact this is not necessarily true, since the security layer is quite independent of the communication type and protocol.

Realistically, however, most smart-card readers today outside public transport use a contact interface only. This means that any card scheme that seeks to include both transport or ID and another application, such as payment, must have both contact and contactless interfaces.

A dual-interface card may offer the same set of applications across both interfaces or – more often – each application is restricted to one interface (with possibly some common applications that may use both).

3.2.4   Dual-chip cards

It is also possible to design cards with two chips: one attached to the contact plate and the other to the contactless antenna. This provides a secure method for ensuring the separation and security of the two applications, but it prevents the two applications from collaborating. Such cards may be thought of as multi-application from a marketing point of view, but from a technical perspective they are effectively two cards.

3.3   Readers and terminals

A smart card must be inserted into a reader or terminal to operate; the functionality always comes from the combination of card and reader. In just the same way as described for cards, readers may be 'multi-functional' in several different ways; it is relatively easy for a terminal to carry several programs that can be selected according to the card that is inserted, the context of a transaction, or even from a keyboard or operator command.

Chip-card readers are more accurately known as card-accepting devices (CADs), since they can write data to the card as well as read from it. They are also sometimes called read-write units (RWUs) or interface devices (IFDs).

Many readers are purchased as part of a terminal meeting one of the common application standards, such as EMV, GSM or Calypso. Terminals for multi-application card systems may need to meet more than one application standard, and this means either two or more sets of certification, or in some cases a tailor-made design that can meet both sets of requirements.

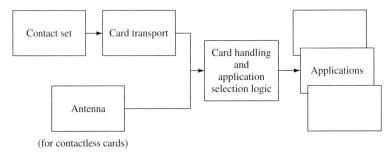

(for contactless cards)

Figure 3.4 Logical structure of chip-card terminal

Figure 3.5 Insertion reader (© Magtek, Inc.)

3.3.1   Components

Smart-card terminals come in many shapes and sizes; in all cases, however, they will contain a card interface (either a contact set or an antenna) and some electronics to manage the power and communications. The software and logic for the application itself may be in the same or a separate physical unit as the reader.

Linking the two (see Figure 3.4) is the application selection logic, which plays a key rôle in any multi-application system, and which will be described in Chapters 8–11.

3.3.2   Contact sets and card transport

The core of a reader for a contact smart card is the set of between five and eight contacts that provide the electrical connection with the card. This may form part of an insertion reader (see Figure 3.5) or a motorised reader (see Figure 3.6).

Insertion readers are much cheaper; they have few moving parts and require little power. In many applications they are the obvious choice.

Where it is important for the application to control the card throughout the transaction (for example, where there may be a requirement to capture the card or the cardholder might be tempted to withdraw the card early) a motorised reader is preferred. Motorised readers are often preferred for unattended operation, because they offer more guidance to the user, and they can also incorporate a shutter covering the slot, which reduces the risk of vandalism. However, motorised readers are more expensive and have many more moving parts, so that it is much more costly to achieve high reliability.

Figure 3.6 Motorised reader (© Magtek, Inc.)

3.3.3   Terminals

Most smart-card readers are bought as part of a larger terminal unit. The requirements for terminals vary greatly according to the sector and are discussed in more detail in Chapter 7.

3.4   Standards

Standards are very important to the smart-card industry, and as we will refer to them frequently throughout this book, here is an introduction to some of the most commonly used standards. A more complete list of standards is given in Appendix C.

3.4.1   Physical and magnetic stripe

Most identification cards follow the **ISO/IEC 7810** standard – the shape and size of card we are all familiar with. Cards with a magnetic stripe (usually known as magstripe cards) also follow **ISO/IEC 7811**, which defines the location of the magnetic stripe and the way data are encoded onto it, in three tracks. Nowadays, most large-scale mag-stripe projects use high coercivity stripes (ISO/IEC 7811-6), which are more reliable than the older low coercivity stripes because they cannot be affected by common permanent magnets and stray fields.

The numbering system for card issuers is also defined in ISO standard **7812**, while **ISO/IEC 7813** defines a financial transaction card (bank card).

3.4.2   Smart cards

This tidy, single hierarchy becomes much more complicated when we move to smart cards. Again, there is a family of ISO standards: **ISO/IEC 7816**, which covers normal, contact-based cards. However this runs to 15 parts: perhaps the most important are **Parts 1 and 2**, which cover the contact plate layout and usage; **Part 3**, which defines the electrical and protocol interfaces and how they may be selected; and **Part 4**, which defines how an application may be selected. Part 4 is particularly important for some aspects of multi-application usage, and so is covered in more detail in Chapter 8. Several other parts seek to standardise data formats, commands and other functions; inevitably some will fall by the wayside but others may become increasingly important, particularly for inter-industry use where ISO standards are the one universally accepted source.

However, alongside ISO 7816 there are also several standards for contactless cards:

- **ISO 10536** is a standard for close-coupled contactless cards (up to 10 cm), and is not widely used.
- **ISO 14443** (proximity cards) covers the two most common contactless card types used today (known as type A and type B). Several other proprietary standards were proposed for inclusion in this ISO standard but were not accepted.
- **ISO 15693** defines a longer-range card, known as a vicinity card.
- **ISO 18092**, published in 2004, defines the 'near-field communications' (NFC) interface for communications between devices such as mobile phones and terminals; it leans heavily on ISO 14443A and Sony's FeliCa^{TM}, another proprietary system for contactless cards, which is widely used in the transport industry in Asia. The specific advantage of NFC is that by using the short-range (non-propagating) characteristics of the antenna, it not only gives very predictable performance with a low power requirement, but also ensures that devices can only communicate when in very close proximity.

This is still a developing field and it is quite likely that there will be further progress (such as higher bandwidths and greater security) over the coming years.

3.4.3   Application standards

All of these smart-card standards define the lower (electrical and data-link) levels of the card-to-terminal interface. Above these levels, international standardisation has been much weaker, and inter-sector co-ordination almost non-existent. Each of the main application sectors has developed its own standards, for example:

- **EMV**: Between 1993 and 1996 the major credit-card schemes (Europay, Mastercard and Visa) developed a further set of specifications based on ISO 7816, but covering the core functions of a bank card in much more detail;
- **ETSI**: The European Telecommunications Standards Institute has been responsible for a set of standards that cover smart cards for use in public and cellular telephone systems, in particular GSM, 3GPP and its smart-card platform, all of which are covered in Chapter 14;
- **ITSO** and **Calypso**: The public-transport sector, particularly in Europe, has sought to achieve interoperability between ticketing systems and modes of transport by developing specifications for a smart-card-based ticket and payment structure;
- **IATA**: The airline industry association has developed a specification for an electronic ticket.

Each of these standards will be described in the relevant chapter within Part III. However, even this short list gives a hint of the challenges that may be faced by any operator or developer of a card system that seeks to include more than one sector in its scope.

3.4.4   Testing standards

Also important are the standards for card reliability and for testing cards. These include ISO 10373 at the physical, electrical and radio-frequency levels, but each sector also has testing processes relevant to its own application standards.

3.5   Smart-card manufacture and supply

The smart-card industry comprises a long and complex supply and value chain – see Figure 3.7.

On the card's side, we start with semiconductor manufacturers. These are nearly all very large companies with a wide range of products; a very large scale of investment (several billions of dollars) is required both for setting up the manufacturing facilities and for designing the products. Smart-card chips, therefore, compete for manufacturing facilities with other products, such as dynamic memory (DRAM), graphics processors and PC chips, which can earn much higher margins for the manufacturer. Smart-card chips cover a wide range from the lowest-end memory chips, costing a few cents, to relatively high-value crypto-processors, which the manufacturer sells for up to $10.

As in other parts of the semiconductor business, some new 'fabless' manufacturers are now specialising in design, and have their products made by others.

Most smart-card chips are sold to card manufacturers, some of whom evolved from magnetic stripe card manufacturing and encoding. Their speciality is in handling

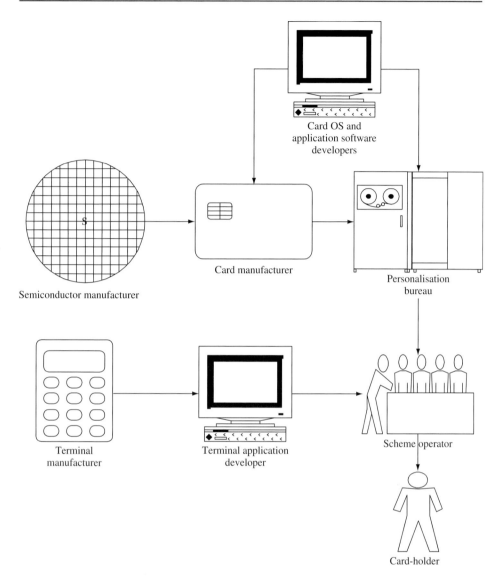

Figure 3.7 Smart-card supply chain

the plastic and the process of embedding the chips; the range of processes and materials used is surprisingly wide and is discussed in more detail in references [1], [2] and [3]. The security of manufacturing plants is very important, and those that want to supply government or financial cards must undergo regular inspections and certification.

   Most smart-card manufacturers today were either set up as, or have evolved into, integrated operations that include not only the embedding but also the development of smart-card operating systems and on-card applications. This allows an integrated view of security, and also enables them to specify products to the chip manufacturers in more detail.

All card manufacturers also undertake card personalisation: the process of storing data on the card. This is often done in two phases: storing the programs and standard data for the card issuer, and adding the data that are specific to the card-holder. In addition to the card manufacturers, there are several specialised bureaux and service companies that offer personalisation and key management only, often in conjunction with other card management services.

Cards are then delivered to the scheme operator or despatched directly to the card-holder.

The supply chain on the terminal side is shorter, although there is growing specialisation among terminal application developers, many of whom will seek to develop applications that can be deployed on several different terminal platforms. This is often more efficient, but it can conflict with the need for integrated security, and independent software companies are often hampered by a lack of development tools designed for external use.

In most industries, there is a single scheme operator, who is responsible for the relationship with all customers, for card issuing and for handling transactions. In the financial industry, described in Chapter 15, these rôles are divided between a card issuer, a merchant acquirer and a card scheme. In all industries, though, it is common for activities such as network management or card distribution to be subcontracted to specialist organisations.

We will see in Chapter 12 that with the growth of multi-application smart cards, card management is becoming a more complex and important part of the value chain. Suppliers and operators of card management systems will play an increasingly important rôle in the years to come.

3.6   References

[1] Hendry, M. *Smart Card Security and Applications.* 2nd edn, Artech House 2001
[2] Zoreda, J. L. and Otón, J. M. *Smart Cards.* Artech House 1994
[3] Haghiri, Y. and Tarantino, T. *Smart Card Manufacturing: a Practical Guide.* Wiley & Sons 2002

Part II

Technology

4 Biometrics

We must now digress from our discussion of the technology of cards and readers to cover another two sets of technologies that have a major impact on cards, and are likely to grow in importance in the coming years: biometrics and cryptography. This chapter will cover biometrics and Chapter 5 will address cryptography and the security of cards.

In most smart-card applications, the card is associated with a person; it represents a key to that person's details in a database. It is, therefore, very important to be able to identify the person who is using the card and to ensure that he or she is the person whose details are being unlocked. Often (for example in an access control or passport application), this is the main purpose of the card or application. In other cases the purpose is to allow access to data stored about that person.

In most card-based systems we need to do this automatically, although sometimes there is a human element as well (for example, inserting the card could call up an image of the card-holder, which can be checked while the user is entering his or her password or verifying a fingerprint).

Many people feel that a human check is better than any automatic check, however this is definitely not true when the population being checked is large. In this case it is not wise to rely on manual comparisons only – the level of discrimination that can be achieved in any manual check of strangers (whether comparing faces or signatures) is so low that the checker is more likely to challenge genuine users than detect impostors. A good test is: 'Would the person doing the checking know everyone they are checking by name?' If not, an automatic check, or a combination of the two, is preferable.

A specific advantage of a card-based system is that it rarely needs to *recognise* the person (do I know this person?); much more often the task is to *verify* the person's identity (is this the person he or she claims to be?). This is much easier than recognition, which involves lengthy database searches.

4.1 Identification requirements

4.1.1 Passwords, tokens and biometrics

Nearly all identification methods are based on something you *know* (a password or PIN), something you *have* (a card or token), or something you *are* (a behavioural or physical characteristic, known as a biometric).

Each of these has its advantages and drawbacks, and each may be appropriate in different circumstances. PINs and passwords are easy to check – they are either right or wrong. Although passwords are often forgotten, written down or gained by subterfuge, they do have the advantage that they can be transmitted over communications networks and they do allow delegation. Cards and tokens need a device to read them; they must be robust and not easily copied – since they can be lost, stolen or borrowed, they are rarely an adequate proof of identity on their own.

Biometrics, which require special hardware and software, yield a confidence level rather than a yes or no answer. And they are often inflexible – few biometric traits can be verified remotely, nor do biometrics allow delegation. But they are, for the most part, secure against counterfeiting and theft. Systems designers must be careful to choose the right form of identification for each situation.

This is a challenge for a multi-application system, where the ideal form of identification may vary from application to application, but one of the advantages of smart cards is that they allow these methods to be combined. Using two factors (e.g., a card and a biometric) is always stronger than any one factor, and two-factor authentication is often required as a policy where even moderate levels of security are required.

By definition, when the card is used there is a token. The card may, though, store a password or PIN as well as a fingerprint or other biometric. These may be used in situations requiring different levels of security, or simply according to the checking methods available. For example, a bank might offer a fingerprint reader at its own ATMs, with a high level of functionality and access to the account, but if the card is used at another bank's ATMs then the PIN would be used, for cash withdrawal only.

There is growing demand for biometrics in identification and identity verification, fuelled by increasing requirements for personal and IT security. There is a general perception that biometric-based identification is always more secure than other forms. This is not a safe assumption; although it is possible to design very secure systems using biometrics, performance issues may limit the security in any given application or environment.

4.1.2 Performance

To be usable for a large-scale system, a biometric must meet three criteria:
- It must enable virtually everyone to enrol;
- It must be reliable and quick in operation;
- It must reject impostors while accepting valid users.

Several technologies fail even at the first step, either because they rely on characteristics that not everyone has (e.g., a good clear fingerprint on a particular finger), or because the characteristic is not sufficiently repeatable in all people (e.g., signature dynamics).

There are also user-acceptance factors to take into account: whereas in a prison or other high-security environment it may be acceptable to ask users to place their eye right up to a camera lens, this is not likely to be workable for customers in a shop. In Japan, hygiene considerations mean that many customers would not be keen to place their finger or palm on a pad at a public terminal.

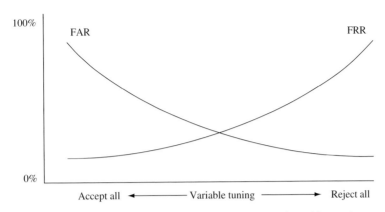

Figure 4.1 False acceptance rate vs. false rejection rate for a biometric

Most of the technologies described later in this chapter have been developed to the point that terminals and software are now very reliable in operation, but some are dependent on the environment – for example, almost any technique that relies on a camera (facial recognition, iris scanning and even some fingerprint techniques) is affected by the ambient light levels; they may not work consistently at night or in bright sunlight. Most biometrics work quite well inside buildings, but many fail at the critical stage of letting people enter the building!

The other key characteristic is the ability to discriminate between valid and invalid users. This is usually measured by the false-acceptance rate (FAR): the percentage of impostors accepted by the system, and the false-rejection rate (FRR): the percentage of valid users rejected. Most systems can be tuned to provide either sensitive detection (low FAR but high FRR) or coarse detection (low FRR but high FAR) – see Figure 4.1.

4.1.3 Interoperability

In a multi-application system, interoperability may also be important: systems and products from different manufacturers must be able to work together. This has traditionally been a major obstacle, as each manufacturer developed proprietary algorithms and template structures. There is growing support for a general Biometric Application Program Interface (BioAPI^{TM} [1]) that, in principle, allows both application developers and biometrics providers to work to a common interface. Many suppliers support this API.

In the United States, the National Institute of Standards and Technology (NIST) has also supported the development of a Common Biometrics Exchange Formats Framework (CBEFF) – [2] – that allows the exchange of information on a biometric in a manufacturer-independent format. NIST is seeking ISO adoption of this framework.

There remains, however, a confusing array of techniques and manufacturers' claims, which has probably constrained the application of biometrics in many commercial applications. Users will not select any biometric until it is clear which biometric measure is most effective, and, within that measure, which solution is the most durable and future-proof.

4.1.4   Procedures

We can identify three stages in a biometric process: enrolment, verification and update.

At the time of initial enrolment, the characteristic is measured and a template is produced. Most templates are algorithmically derived representations that do not need to store all the measurements, but only some key features; if these features provide enough discrimination, the template may be very small: a few tens of bytes in some cases. The measurement is usually carried out three or more times to ensure accuracy and consistency.

Users are verified by re-measuring the characteristic and comparing it against the template. They must often be helped to provide a consistent measurement, and even then some tolerance is required.

Many characteristics change with time and so it is also important to make provision for updating the template, either continuously or when pre-set tolerance levels are exceeded.

4.2   Biometric technologies

Biometric techniques have been developed to measure hundreds of different physical and behavioural characteristics, ranging from fingerprints and face recognition to DNA matching. But the range of techniques usable in the field is much smaller; they include:

- **Fingerprint** or **thumbprint**: this is one of the oldest forms of identity check; although it was formerly associated with criminality, it is now becoming much more widely used and more acceptable to the public. Fingerprint scanners can be built into keyboards and mice (the manuscript for this book is being typed on a PC fitted with a fingerprint reader). There are two major groups of fingerprint-recognition product: those that use a photographic scan of the whole print and those that rely only on the minutiae – the key points such as intersections or ridge endings. The major advantage of the second is the small template size (100–500 bytes), which is attractive in a smart-card application.
- **Face recognition**: high-speed image processing is making automatic face recognition increasingly viable. It is often used by police and at borders, where despite a relatively high FAR it is an effective deterrent and a useful screening tool. As with the fingerprint, there are two types of template: the full image and the key characteristics (measurements of distances between the eyes, nose, mouth, etc.). It is likely that with the use of facial images in passports and ID cards (see Chapter 17), this will become a widespread form of biometric, although the template sizes are

larger than for fingerprints, placing a premium on fast card-to-reader communications.

- **Hand geometry**: measuring the relative lengths of different fingers and joints scores highly for ease of use, since the hand is large and can be placed very consistently using pegs and guides. Templates can be very small but the readers are large and not suited to some environments. Hand geometry is often used for employee clocking-on, and it has also been used at major sporting events, such as the Olympics.

- **Finger-vein (and palm-vein) patterns**: these newer techniques [3] use a 'near-infrared' light source to illuminate and scan the veins in the finger or palm: see Figure 4.2. The finger-vein reader is as small as most fingerprint scanners ($39 \times 34 \times 15$ mm) and as usual, the raw data are processed to yield a small template. Because the device is measuring the light absorption of haemoglobin, it will only work with a live finger or hand, but it could also work through a translucent glove, which would be seen as an advantage by Japanese users. This technique is being tested in several thousand ATMs in Japan and, if successful, could be quite widely adopted.

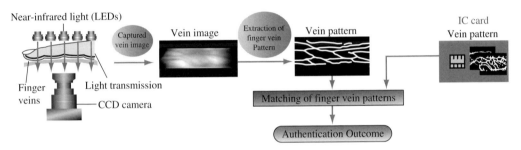

Figure 4.2 Finger-vein verification (© Hitachi, Ltd)

- **Retina scan**: retina scanners measure blood vessel patterns on the back of the eye using a low-power laser and camera. The user must place his or her eye close to the camera for a focused image, which is a concern to some people, and these patterns do change with age. As a result, retina scanning is now giving way to the less invasive iris scan.

- **Iris scan**: this technique [4] analyses the tissue network surrounding the iris of the eye (see Figure 4.3); the ideal scanning distance is about 30 cm from the camera, and it can be used by spectacle and contact lens wearers. The most widely used and patented method scores highly for accuracy and repeatability, with a moderate template size.

- **Automatic signature verification**: this uses the dynamics (pressure, speed, etc.) of signature, rather than the final shape. However, the shape is also captured, which allows a permanent record of the signature to be stored. This is often attractive as an addition or upgrade to a visual signature-verification system.

- **Voice recognition**: this form of biometric is very easily acceptable to customers, and has the advantage that it can easily be combined with a password or passphrase (the user speaks the phrase). The system is most dependent on the size and shape of the 'voice box' in the throat and the way sounds are produced, and so it is less open to

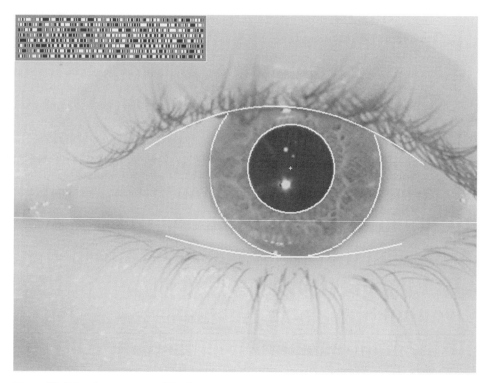

Figure 4.3 Iriscode (courtesy of Prof. J. Daugman)

impersonation than might be thought; however it is prone to high levels of false rejection because of colds, background noise and the inconsistency of the transducers and transmission system. Voice recognition works over telephone lines and has been used by Sprint since 1993. But it requires continuous adaptation and updating and is not generally used in smart-card systems.

Longer lists and more detailed descriptions of biometric techniques can be found in any biometrics textbook, e.g., [5], or on many websites such as the Biometrics Consortium [6], Avanti or the International Biometrics Foundation [7].

4.3   Biometrics in cards

When a card is used as part of a biometric system, there are several tactics that may be adopted for verification:

- For small systems, the card may point to a database record where the template is stored: all measurement and verification is then carried out by the central system;
- The template may be stored on the card but transmitted to the host or terminal for verification purposes;
- The measurement and extraction of the data for comparison may be carried out by the terminal, but the actual comparison takes place within the card; or
- The raw data are passed to the card for matching.

Each of these represents a different security model (who trusts whom?) and there is no universally correct answer. Whenever the template or reference data are passed over a communications link, there are opportunities for the data to be copied and misused. In many personal data applications, it is very important for the user to remain in control of his or her own data; this helps to protect the user's privacy as well as offering portability. For national ID card systems, for example, this is a strong (but often underplayed) argument in favour of storing the template on the card.

For a multi-application system, particularly an open system with many card issuers and acceptors, maximum flexibility is achieved when the card stores not only the template but also any special algorithms that may have been used in its creation, so that the same algorithms may be used during verification.

Traditionally, card systems have sought to use biometrics with very small template sizes: this affects not only the storage requirements on the card, but also the verification time, since either the reference template or the extracted data must be transferred from or to the card during verification. This worked in favour of hand geometry (typically 9 bytes) and retina scans (35 bytes).

As we saw in Chapter 3, for a basic contact card running at 9600 bps, even if the chip has sufficient memory to store the 1–1.5 kB needed for a full-scan fingerprint image, the data exchange between card and reader is likely to take at least 2 seconds in that case. With growing memory sizes and faster communication speeds, these requirements are no longer as critical, and adequate performance can be achieved with full-scan fingerprint images and even face recognition (4 kB).

We will see in later chapters that the security of the template and associated algorithms is very important: we must avoid the 'yes-card' attack, in which the card always responds positively to a request for verification. There must be a suitable level of authentication between card and reader, or between card application and reader, and templates should also be cryptographically signed by the application that generated them. These requirements will be discussed further in the next chapter.

For a card-based system the need for updates to the template poses a dilemma: can the template stored in the card be updated during a verification transaction or does this require some host system intervention? Some cards with short issued lives (up to 2 years) may be updated only when they are re-issued; otherwise there should be some provision for updating the template, either 'on the fly' or offline.

Cards should always store a record of enrolments, and successful and failed verification attempts; it is also an advantage if they store the confidence levels achieved at each verification, since this can show when a feature is going out of tolerance or a user is naturally inconsistent. This information may be made available to applications after verification.

4.3.1   On-card data capture

One potentially very interesting development is the concept of having the sensor and template extraction in the card as well as the template storage and comparison engine. This would allow the card to perform the whole operation securely, and would boost privacy as well as preventing any part of the template or image from being copied. However, it does make the system as a whole more prone to 'yes-card' attacks and so may be more appropriate to personal data storage applications than for corporate or government use.

In the past, chip manufacturers Infineon and ST Microelectronics both developed the combination of an ultra-thin fingerprint scanner and microcontroller, capable of fitting in the thickness of a card, but found that the template extraction required specialised or very fast processors, which demanded too much power for normal card use. More recently, US company Smart Metrics has combined these with a battery in the card, which may make the concept usable in some environments.

4.4   References

[1] BioAPI™ Consortium. www.bioapi.org

[2] National Institute of Standards and Technology (NIST). *Biometric Technologies: Helping to Protect Information and Automated Transactions in Information Technology Systems.* ITL Bulletin September 2005. http://csrc.nist.gov/publications/nistbul/bulletin-Sept-05.pdf

[3] www.hitachi.co.jp/Prod/comp/fingervein/global/

[4] Daugman, J. *Introduction to Iris Recognition.* www.cl.cam.ac.uk/users/jgd1000/iris_recognition.html

[5] Bolle, R. *et al. Guide to Biometrics.* Springer-Verlag 2003

[6] *An Introduction to Biometrics.* www.biometrics.org/html/introduction.html

[7] www.ibfoundation.com

5 Security and cryptography

Another science – some would say art – that is very important to smart cards is cryptography. Cryptography is an essential part of many of the security functions for which smart cards are used. This chapter can only give an overview of the issues that are relevant to smart cards, and readers seeking a deeper understanding of algorithms and cryptography generally are referred to the further reading suggested in Appendix B.

5.1 Cryptography

5.1.1 Algorithms

Modern cryptography combines algorithms (mathematical transformations) and key management techniques to secure data in many different ways. The main algorithms used change only very slowly, since only thoroughly tested and well understood algorithms are used for important security functions. People outside the security industry often feel that a newly developed or secret algorithm should be more secure, but the history of cryptography has shown that only a very few algorithms remain unbroken after many years. Nearly all others succumb sooner or later to some easy attacks – once an attack is known the algorithm is useless.

Algorithms are divided into two groups: *symmetric* algorithms (like the Data Encryption Standard ANSI X3.92 or its more modern and stronger replacement, the Advanced Encryption Standard FIPS-197 [1]) use the same key for encryption and decryption. *Public-key* algorithms (such as RSA [2]) use a different key for encryption and decryption: the owner keeps one key private while the other is published.

If Romeo wants to send a message to Juliet, he can look up her public key in a directory and encrypt the message using that key. He is then confident that only she can decrypt the message. He also takes some key fields from the message (known as the *digest*) and encrypts that with his private key. Juliet can look up his key and check that the message really did come from him (see Figure 5.1).

Generally, symmetric algorithms are easier to implement but it is more difficult to manage keys, whereas the public-key half of a public–private key pair can be relatively widely distributed. However, the private half of the pair must be kept very secure, as must the *secret key* used in a symmetric system.

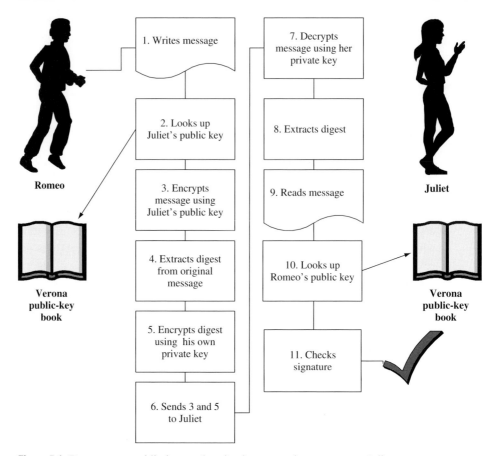

Figure 5.1 Romeo uses public-key authentication to send a message to Juliet

Encryption algorithms are often further divided into *stream ciphers* and *block ciphers*: as the name suggests, stream ciphers are more efficient at encrypting streams of data of indeterminate length while block ciphers can only handle delimited blocks, often only the same length as the key and rarely longer than 1 kB. Most smart cards are used for single transactions, with defined message types, and so block ciphers are more widely used. Even where smart cards are used in stream-encrypting devices (such as encrypting modems) they more often supply an initial key rather than perform the data encryption on the fly.

Most smart cards can perform DES encryption relatively efficiently, since this is a logical operation that does not need any special hardware or arithmetic type. For almost any other algorithm, the speed of encryption, decryption and verification is greatly improved by having a cryptographic co-processor on the card. These co-processors include special hardware and software that allow them to perform variable-length arithmetic and shift register operations, as well as low-level software that provides single calls for encryption, decryption and verification using the most widely used algorithms.

These will, as a minimum, include DES with single, double and triple-length keys, AES and RSA. Many will also include a variety of hash functions, Diffie–Hellman authentication and elliptic curve encryption/decryption [3] – the latter is seen as particularly useful for smart-card cryptography as it uses relatively short keys and fast encryption for a given strength compared with other public-key algorithms.

5.1.2   Key management

Smart cards play an important rôle in many cryptographic applications, because they allow private and secret keys to be stored in a section of memory that cannot be accessed from outside the card: only an application on the card with the appropriate permissions for that memory area can access those keys. (Later in this chapter we will consider some potential attacks on keys stored in this way.)

Key management techniques determine how keys are generated, stored, transmitted and refreshed. Keys used for different purposes must be protected in different ways, and this often involves a hierarchy of short-term and long-term keys, system keys and device keys, encryption keys and authentication keys. Key management is governed by security policies and security zones (within which certain protections are assumed to exist). Each sector has specific techniques for setting and managing security policies and often, as we will see later in this chapter, specific standards.

The file structure used by all ISO 7816 smart cards (i.e., all microprocessor cards) provides a basic key-management structure and functions. Most wired-logic cards have a similar hierarchy. Smart cards with cryptographic co-processors can sometimes make use of special operating-system commands for key management, such as key encryption.

The main multi-application card operating systems, which are described in Chapters 9–11, provide a much higher level of key management and this is often supported by special host-system functions.

5.2   Security services

IT security professionals classify all security requirements into five sets of security services: confidentiality, integrity, authentication, non-repudiation and availability. They recognise that there must be many layers of security in most systems, and that protection such as physical security often has an important rôle to play alongside technical measures. Smart cards use cryptography to help provide each of the five recognised security services:

5.2.1   Confidentiality

Confidentiality of data or messages is ensured by *encryption* and *access control*.

Strong encryption involves the use of a published and well tested algorithm, which may be either symmetric or public-key. Symmetric algorithms are more commonly used for encryption functions, because they are simpler to use and easier to implement in hardware – the fastest smart cards can perform single DES encryptions in a few micro-seconds, compared with tens of milliseconds for RSA at even very short key lengths.

Key length is the other determinant of encryption strength: single-length DES (with a 56 bit key) is now rarely used in security-conscious applications, but triple-DES with two keys (with an effective key length of 112 bits) is very widely used in banking and forms the basis of many functions in the EMV financial smart-card standard. Three-key triple-DES (where a different 56 bit key is used for each phase of encryption) is sometimes used today and may rapidly become more common. For RSA, key lengths for most commercial applications today are in the range 896–1152 bits (112–144 bytes), but most smart-card crypto-processors are designed to handle up to 2048 bits.

Symmetric keys must be distributed securely to all the units that will access infor-mation, while for public-key systems the private key must be restricted to one logical entity only while the public key can be distributed at much lower risk (although the difficulty of key management in public-key systems should not be understated). In some cases, a key for a given card or operation will be derived from some card or current data, using a master key.

The main function of the smart card in providing confidentiality is to store and transmit keys securely: once a key is stored in the card's secret area, it cannot be accessed from outside, but card applications may use it to encrypt or decrypt data.

Control of *access* to each application and its data is by authentication and access control lists – users and user groups are defined and given certain access rights.

For multi-application cards, care must be taken to limit the scope of each key used and, in particular, to limit the use of master keys to a single function. The key lengths chosen must be appropriate for each application – shorter keys give better performance but less security. The choice of algorithms will often be determined by the standards prevalent in each sector: for example 3DES in banking or A5 [4] in telecoms, so multi-sector cards that must support multiple algorithms may require the use of a crypto-processor where none would be needed by either application separately.

5.2.2   Integrity

Integrity checking ensures the *completeness* of messages or data, and alerts the system if they have been tampered with. It is often combined with authentication and implemen-ted by 'signing' the data or message. Some sectors (particularly government sectors) have very high requirements for integrity-checking of stored data and, therefore, require large amounts of duplication and redundancy (additional bits or fields that help to check the correctness of the main information). This increases the amount of storage required and the time needed for a transaction to exchange data with a reader.

As well as the integrity of data, I should mention the need to ensure the integrity of applications stored on the card. Multi-application card operating systems ensure this through their control of application loading, but for other cards the same effect can be achieved by storing a 'hash' value derived from the actual code (a hash value should always change if even one bit of the data used to create it changes, but is small enough that there is no way to derive even part of the original from the hash value). This hash value can be verified and used as part of the card authentication process, thus ensuring not only that the card is valid but also that the application code has not been altered.

5.2.3 Authentication

One of the main strengths of smart-card-based security schemes is their ability to provide strong and auditable authentication. Using secret or private keys stored in the card, a host system can authenticate the card (i.e., show that it is a valid card issued by a valid issuer) while the card can authenticate the card-holder (i.e., verify his or her identity). As discussed in Chapter 4, this 'two-factor' authentication adds a dimension of security and is generally the most appropriate for commercial applications invol- ving moderate or large numbers of people.

Standards of authentication vary widely from sector to sector, and this is an area where any multi-sector scheme will inevitably have to make compromises. For small- scale, closed-user-group schemes, symmetric cryptography is usually used as the basis for authentication: it is simpler to set up and offers much faster verification times. It is also a viable option for any system where the authentication takes place fully online between the card and the host.

For large-scale and open schemes, however, and particularly where a terminal will play a rôle in the authentication process, symmetric cryptography imposes significant restrictions and management issues. A security application module (SAM) – a smart- card microcontroller, usually in ID-000 format, embedded in the terminal – will almost certainly be required, and the whole terminal must be tamper resistant, since the processes of card and terminal authentication depend on the secrecy of the keys used. If a terminal or SAM is stolen, the whole scheme may be compromised. Public- key cryptography avoids or reduces these issues and is, therefore, preferred for large- scale public schemes where the security of the infrastructure cannot be guaranteed.

This may, in turn, lead to performance issues or an increase in infrastructure cost for another part of the scheme. For example, in Chapter 16 I describe the cards and standards widely used in public transport, which use symmetric cryptography and demand very rapid transaction times. If this is to be combined on a card with a payment application such as EMV, which uses public-key cryptography, the two applications are unlikely to be able to co-operate, which may lead operators to question the logic for combining them on a single card.

5.2.4 Non-repudiation

Non-repudiation means that the user cannot claim that a transaction did not take place. It follows from authentication, provided that the key used for authentication is unique to each user – it must be safe to assume that no two users have the same key. However, it must also be backed up by rules, and in many cases a legal framework. Customers must accept (in their minds as well as in law) that no-one else could use their card or generate the same digital signature.

The traditions and expectations of users in different spheres of activity have a large impact on the types of systems needed to manage disputes. In a school, for example, authority is respected and disputes have a short path. If a payment application is added to a school card, the rules may change to those of a financial institution, where (in many countries) the law provides a range of alternative, and often lengthy, processes for resolving disputes.

5.2.5   Availability

Availability is an important security service, but is often overlooked in commercial applications alongside the previous four – whereas, for example, nuclear-power-station control designers put it at the top of their list.

Smart cards help to improve availability in systems by allowing more *offline functions*, so that the system is less dependent on the reliability of external networks and other infrastructures. In some businesses, offline working is impractical, but this should not prevent other applications that share the card from working offline if they can. Although smart cards should not be used as general-purpose computers, they can often contain sufficient data and logic to make simple decisions on behalf of the application owner.

Compared with many legacy systems (for example, magnetic stripe cards) smart cards can usually improve availability by offering higher reading reliability and a longer life.

Smart cards are used in some environments for data collection and storage, and they are often used to store keys (for example, in Chapter 14 I discuss their use in satellite television receivers). However, there is an operational conflict between the needs of applications that are essentially transactional (the card is used for a short time while the transaction takes place) and those that require the card to remain in the reader for long periods of time. In the second case, the card is often left unattended, which negates one of its benefits as a transactional token. So it is likely to be difficult to combine these two groups of application.

5.3   Smart-card attacks

Users must be aware of some potential attacks, usually involving the destruction of the card (see, for example, [5]). Card manufacturers are not always able to eliminate these completely, but can make them very difficult and costly, and the system design must limit the impact of the keys stored in any one card being broken or accessed.

5.3.1   Trojan horses

At the design stage, a programmer can insert functions that allow keys or other sensitive data to be read. Often these are not deliberate attacks, but test functions or other 'undocumented features' that were never intended for production use. Such functions must never be allowed in secure-systems design; they can be avoided through the use of structured programming tools and distributed controls (no one person may both create and install a software module, for example).

Similarly, hardware test features (which are sometimes necessary) must be documented and destroyed after use. Design analysis tools can certainly prevent these in legitimate designs but there is a risk that a rogue design or chip could be introduced during the later stages of card manufacture; this shows the need for multiple layers of control, which is a common theme of secure-systems design.

5.3.2   Counterfeiting

Although supplies of smart cards from reputable manufacturers are strictly controlled, there are many ways of obtaining cards from other sources (just look at the internet pages devoted to cable television decoding cards, and you will see that only a small proportion of these are legitimate).

If the data can be read from a real card and copied into a counterfeit card (for example an expired card or a blank card of a similar type), then it may be possible to carry out transactions. More secure cards and systems make use of the secret data areas on a card that cannot be read from outside, for example by including in the transaction message some data that have been signed using a unique key stored in this way.

5.3.3   Microprobing and electron microscopy

It is possible, using concentrated acid, to dissolve the protective layers covering the chip. Tools that exist in many semiconductor laboratories can then be used to probe or inspect the conducting layers underneath. These attacks have become extremely difficult as chip geometries have grown smaller (the probes are larger than the lines on the chip), and as the card is destroyed in the process there is only value in this attack if the secret obtained helps to make more than one new card. For this reason symmetric keys that are valid system-wide should not be stored on the card.

5.3.4   Environmental attacks

It is sometimes possible, by manipulating the temperature, voltage or clock signal, to cause a failure in the cryptographic processing; this can help to determine intermediate results and hence to extract parts or all of a key. There is a strong element of chance in these processes, and chip designers are generally able (by using temperature and voltage sensors, free-running clocks, etc.) to prevent the system from giving any output at all in the event of attacks like this.

5.3.5   Differential power analysis

A very powerful set of attacks, discovered by Paul Kocher and colleagues [6], is the analysis of the chip's power consumption as it performs cryptographic tasks and key-searches. With some knowledge of the algorithms and hardware characteristics, keys and secrets can be extracted during normal processing, simply by inserting a probe into the power connection. This attack is quite likely to be effective unless the chip and firmware designers have used specific techniques to deter it, such as the introduction of random wait times and null operations. Where security is an issue, projects must insist on a DPA-protected chip.

5.3.6   Yes-cards

I discussed in the previous chapter the advantages of using on-card matching for biometrics (and the same may often apply to PINs and passwords). However, this technique opens up the possibility of creating a special card that always indicates a positive result for the biometric or password match; these were dubbed 'yes-cards' by the French smart-card hacker Serge Humpich, who was subsequently sent to prison for a successful attack (even though his main purpose was to draw attention to the exploit).

The more powerful and reprogrammable the card, the more useful it is for producing 'yes-cards'; however, these attacks can be avoided by using some of the unique card data (and possibly also a hash of the application program) during card authentication.

5.3.7   Message interception

Lastly there remains the possibility that an attacker can intercept messages to and from the card, or even set up a 'man in the middle attack', where the attacker receives messages but alters them before sending them on to the host system. These should be avoided by encrypting critical parts of the communications between card and host, using a secret shared only by the card and its host.

5.3.8   Preventing attacks

We can see from this list that there is a wide range of potential attacks, and no guarantee that more will not be invented during the lifetime of any card system. The more the system depends on cryptography and other purely technological methods for its security, the greater the likelihood that attackers will use 'low-tech' methods: bribing insiders or using confidence tricks to obtain passwords.

Effective security depends on strength in depth: there must be several layers of security, using different techniques at each layer. As a general rule, there should be a means of *preventing* any known attack, a way of *detecting* a successful or attempted attack, and a way of *limiting the risk* if an attack is successful.

5.4   Security standards

To respond to these challenges, each sector has its own standards for security. These are driven partly by business needs (including regulation and contractual requirements) and partly by history: each sector has, over the years, developed an approach that is cost-effective in its specific environment. Smart cards can sometimes be a disruptive technology and allow new techniques to be deployed, but where there is a large body of investment in an existing technique, there is likely to be strong resistance to adopting a new one.

5.4.1   Cards

Given the importance of security in many smart-card applications (often the application is there to provide security for some higher purpose), the security of the card itself is an important factor in specifying and selecting a card. The card manufacturer puts considerable skill and effort into ensuring that the card itself cannot be tampered with, nor data stored in the card be read in unauthorised ways.

To demonstrate that their cards are secure, an increasing number of both chip and card manufacturers subject their products to a security evaluation using the Common Criteria for Information Security (CC), ISO 15408, or their predecessors ITSEC and TCSEC. The CC evaluation offers various levels of security checking, ranging from a simple functional check (EAL1) to an exhaustive verification of all aspects of the design process, including formal models and testing of the specific security features (EAL7). However, Common Criteria evaluations are not yet available up to EAL7 and, therefore, many high-security systems are still certified against ITSEC or TCSEC.

For a smart card, the lowest level that adds value is EAL4 (this is the first level at which code is independently checked) while some cards have EAL 5 +. This is not to say that cards without such a rating are insecure, however the CC rating does show that independent laboratories have rigorously checked the design and may give an added level of comfort in applications requiring high security.

Common Criteria evaluations refer to protection profiles (PPs) appropriate to the type of device or system; there are at least ten PPs relating to smart cards, including several for smart-card ICs alone [7]. To some extent these PPs compete with one another, and manufacturers will choose the PP that they feel will yield the most favourable results for their product; this means that direct comparisons between evaluations are not always possible.

The overall framework of the Common Criteria scheme does prevent inappropriate PPs from being registered, and so products that have certification at the same EAL are broadly equivalent in security level. However, those specifying or purchasing products with a CC evaluation or certification must also be aware of the exact scope covered by the evaluation – is it the chip, the card, the operating system, the application software or the whole unit? It is easy to negate the security of a smart-card platform by using inappropriate software.

Protection profiles for particular applications, such as the German health card described in Chapter 17, for electronic purses and for secure digital signature cards, are often the most useful as they take a more holistic approach. A product could, in principle, be tested against several PPs, although the paperwork burden is likely to make this option unattractive unless sales volumes will be very large.

Many sectors also have certification schemes, usually operated by independent laboratories but under the control of a national or sector authority (for example ETSI in the case of GSM, or a government security agency). The cost and length of time required to obtain a long series of certifications, and to re-certify in case of any changes, can be a serious barrier to the adoption of multi-application schemes.

5.4.2   Terminals and systems

Similar conditions apply to terminals: there are PPs for PIN pads, transactional terminals and terminals used in specific applications.

It is less easy to demonstrate the security of a multipurpose terminal – particularly if there is a general facility to add applications – than for a card or for a dedicated terminal, and in many cases the approach taken is to assume that the terminal itself is not secure. Similarly, unless the network and cabling are entirely under the ownership of the scheme operator, it is not safe to assume that the network is secure, and so many smart-card schemes rely primarily on encrypted and authenticated messages being exchanged directly between card and host. The security standards and certification procedures are, therefore, extended to the whole system, rather than any one component.

This can cause design difficulties for systems that rely on offline transactions, such as public-transport cards and many electronic purses. In these cases we must often rely on a SAM or other secure device that resides in the terminal and protects the integrity of the system and confidentiality of messaging.

5.4.3   Operations and management

Even where the system architecture and technical implementation incorporate a comprehensive and adequate security scheme, the overall system security also depends on policies, procedures, enforcement and physical security.

Again, standards and accepted practice vary widely between sectors, although the ISO standards for IT Security Management (ISO 17799 and ISO 27001) can be applied in any sector and, for any smart-card scheme involving multiple organisations, it should probably be regarded as essential for all involved to follow these standards.

5.5   Requirements definition

Smart cards are acknowledged as a significant tool in the security systems armoury. However, they are only effective as part of a secure *systems design* – we must ensure that the lock is fitted to the right side of the door, and that there are no open windows immediately alongside that would invalidate the security. In the case of multi-application cards, the scope for leaving windows open is increased, and a rigorous approach must be taken to identifying and resolving security issues.

The standard practice for security design is to start with a *risk analysis*: what is it that we are trying to protect and what are the threats? It is always good practice, but for a multi-application card it is essential, to start at a very high level so as to avoid capturing specific, well-understood risks while missing less obvious but perhaps more severe risks, such as a participant going out of business, new legislation, or insider risks that would not exist in the card issuer's main business.

Where a card will cross several applications or sectors, representatives of all the business areas affected must contribute to the risk analysis, since an understanding of each business, its priorities and operational practices is needed. The resulting analysis may be quite complex, as each risk may have different impacts for different players, and each must determine whether it can protect itself by other methods or whether the risk forms part of the smart-card scheme itself.

Established card issuers such as banks and governments have existing procedures for carrying out risk analyses, and for extending the results into a requirements definition for a project. But these procedures may not be adequate for a multi-sector card, while those sectors with no tradition of risk analysis are unlikely to have any benchmarks for assessing risk likelihoods. The novelty of some of the technology introduces further unknowns.

Some key lessons are to:

● Describe risks in language understandable to non-specialists; if it is not obvious what the risk is, then implementors may protect against the wrong thing;

● Where possible, express likelihood in terms of probabilities or incidents per year – this helps to normalise likelihoods and makes it possible to assess the cost-effectiveness of countermeasures.

Even when all countermeasures have been defined, there will remain some risks for which no cost-effective mitigation is possible. All parties must accept this, and understand where they may incur a liability as a result.

5.6   References

[1] *Announcing the Advanced Encryption Standard.* http://csrc.nist.gov/publications/fips/fips197/fips-197.pdf

[2] RSA Laboratories. *What is the RSA Cryptosystem?* www.rsasecurity.com/rsalabs/node.asp?id = 2214

[3] Blake I. *et al. Elliptic Curves in Cryptography.* Cambridge University Press 1999

[4] GSM Association. www.gsmworld.com/using/algorithms/index.shtml

[5] Anderson, R. and Kuhn, M. Tamper Resistance – a Cautionary Note. *Second USENIX Workshop on Electronic Commerce Proceedings* 1996

[6] Kocher, P. C., Jaffe, J. and Jun, B. *Introduction to Differential Power Analysis and Related Attacks.* www.cryptography.com/resources/whitepapers/DPATechInfo.pdf 1998

[7] www.commoncriteriaportal.org/public/expert/index.php?menu = 8

6 Card technology

This chapter picks up the story from Chapter 3 and looks at advances and variants in chip and card technology, particularly those that affect multi-application cards.

6.1 Microcontrollers

6.1.1 Architecture

The microcontroller chip is at the heart of smart-card technology; as we saw in Figure 1.2 an increasing proportion of all smart-card chips use one.

As in all computing, more advanced operating systems and applications are demanding more power from the processor. But while microprocessors for mainstream computing are able to satisfy the demand for more power by increasing the number of transistors packed onto the chip, and hence the heat generated, smart-card chips are limited in both the area of the chip ($25\,\mathrm{mm}^2$ is normally considered the limit for reliability) and the amount of heat that can be dissipated.

A growing number of chips, therefore, make use of reduced instruction-set computing (RISC) cores, which give faster processing for a given power input. Separating the cryptography into its own processor can also help, and it is also more efficient if the input–output is handled by its own processor. Longer word sizes (32 bit words are now the norm in mainstream computers, and 64 bits quite common) are less beneficial in smart cards, and a 32 bit processor does not necessarily give a better speed–power ratio than 16 bits.

Cards can be tailored to the specific applications they will run: Multos cards have for some time been tailored to running the specific code that this operating system generates, and some processors are now optimised to run Java byte code directly.

6.1.2 Feature size

One factor that does reduce energy consumption per transistor, as well as allow higher packing densities, is the feature size or geometry used (the width of the finest line used in etching the chip). The more powerful and large-memory smart-card microcontrollers now use feature sizes of close to $0.13\,\mu\mathrm{m}$ (130 nm). Although processes of less than $0.1\,\mu\mathrm{m}$ are now used for advanced CMOS logic products that can be produced in high volume, such as high-density DRAM memory chips and large microprocessors for PCs, these high-volume (and high-cost production line) products do not incorporate non-volatile E^2PROM memory technology, a key component of smart-card products. This technology is much more complex because of the wide voltage and power requirements and the associated write/erase endurance requirements. So for each step to finer product dimensions, more R&D time is required before an equivalent product can be brought to market.

Although feature sizes do continue to get smaller, the rate of change has reduced and it is clear that the limits of current technology are being approached, particularly for E^2PROM – some breakthrough will be required to drive feature sizes to the next stage. There are theoretical limits: when features become too small, the number of free electrons becomes an issue and control of the current flow through the transistor junction becomes less reliable.

6.1.3   Memory types and sizes

Alongside the demand for computing power runs the demand for increased memory. We will see in Chapter 14 that as far as smart cards are concerned, this demand is being driven by telephone cards (SIMs and USIMs), but other applications such as health cards are not far behind.

For many years, one of the problems of multi-application cards was that each application demanded more memory. Since memory size was a major factor in card cost, it was often cheaper to purchase two cards than a single card with twice as much memory, and much easier to administer. This has greatly changed with the advent of new memory types, and now – for microprocessor cards at least – there is little difference in cost between a 4 kB, 8 kB and 16 kB card; often the card that is being produced in the largest volumes will be the cheapest, not the one with the smallest memory.

Since the early days of smart cards, the most important form of memory has been electrically erasable programmable read-only memory (E^2PROM), which has the advantage of being rewritable in very small blocks, while retaining its memory when the power is removed. However, it requires a large area per bit and has relatively high power consumption, compared with the other components on the chip. For some time it was felt that ferro-electric random-access memory (FeRAM or FRAM) might take its place, but the process for this is proprietary and cannot be integrated with silicon processing (it must be isolated to prevent contamination of the silicon). The promise of equivalent memory densities and reliable manufacturing has not been realised, so few manufacturers have been prepared to invest in it. However, flash memory is being increasingly widely used in a huge array of consumer and embedded applications, and every chip manufacturer now has a flash-memory manufacturing process.

Flash memory has one significant disadvantage: it can only be erased and over-written in relatively large blocks (typically 1 kB); it is, therefore, not well suited to storing working parameters such as transaction counters or stored value. However, its very low power consumption and small footprint on the chip make it very attractive for storing larger quantities of slower-moving data such as a user profile or directory.

The preferred method for many applications, and to make the most versatile chip, is to combine some E^{2}PROM and some flash memory. Chip suppliers can currently fit around 256 kB of flash memory alongside a small amount (e.g., 4 kB) of E^{2}PROM on a single chip without exceeding the power and footprint limitations.

For larger memory sizes (such as the 1GB being demanded by some mobile telephone operators for voice and video clip storage – see Chapter 14) two chips must be used, possibly in conjunction with a 'flip-chip' construction (see below).

6.2   Cards

6.2.1   Materials

Laminated polyvinyl chloride (PVC) remains the dominant material for plastic cards, because of the ease with which it is worked (it is almost the only material that can be embossed). However it is not very durable, while its manufacture and disposal both produce chlorine compounds that are thought to be toxic, and so several other materials such as glycolised polyester (PETG) or even paper are now used. In particular, polycarbonates are more environmentally friendly, offer exceptional strength and durability, and are often preferred for national identity cards and other cards that must last for at least ten years. Glycolised polyester is also quite durable and is sometimes preferred where laser engraving is used for personalisation, since the laser penetrates just the right amount into the plastic, giving a very durable image.

We will see that each sector and application places its own demands on card durability and resistance to temperature and other environmental factors; the range of materials is wide and it is important to select materials that will stand up to all the ways in which the card will be used.

6.2.2   Construction

Smart-card chips are normally connected to the contact pads using wires bonded to each pad; this bonding and the wires themselves are often the weakest element of the structure for mechanical strength and limit the flexibility and packing density of the module.

Many semiconductor devices requiring high connection densities now use a technique known as 'flip chip', [1], [2] or direct chip attach (DCA). Flip chips are placed face-down on the circuit board or other mounting surface (in the case of smart cards, the contact plate). The electrical connection is made without the use of wires, but instead through conductive bumps on the chip's surface; chip and substrate are then bonded together using a non-conductive epoxy (see Figure 6.1). This method of construction offers high densities and improved heat dissipation.

For smart cards, flip chip additionally offers greater strength, a flatter chip surface and the ability to stack two chips on top of one another (hence its particular value for large-memory and multi-application cards). To make this possible, the chip must be

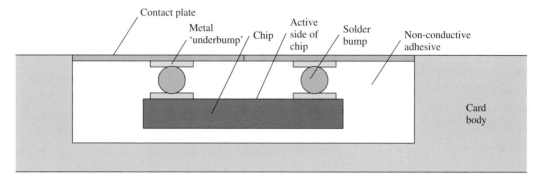

Figure 6.1 Flip-chip construction

'thinned' (have the thickness of silicon reduced by milling or chemical etching) from the 300–400 μm needed for stability of the wafer down to 100 μm or less. Current smart-card chips are thinned down to 180 or 150 μm.

These thinner chips will become more important as packing densities and memory sizes increase, and so will have particular relevance to future multi-application cards. In some applications, stacked chips also offer higher performance (since transmission paths are shorter); in a smart card they offer the possibility of using daughter boards to provide specialised functions in cryptography, cache memory or direct memory access. E-passports also require thinner chips for the optimum combination of strength and flexibility.

The combination of flip chips with thinned dice may also offer some advantages for contactless cards, because there is less silicon surrounding the chip to attenuate the signals.

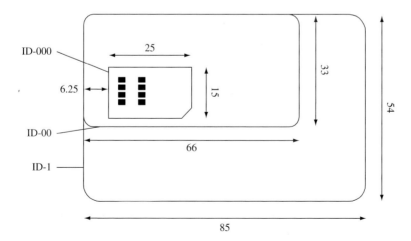

Figure 6.2 ID-1, ID-00 and ID-000 cards (dimensions in mm)

6.2.3   Form factors

A card is a convenient shape and size: not too bulky but easy to handle. There are few applications where it is much too big or much too small, and its thinness makes it very easy for a person to carry several of them at any one time.

However, there are some applications where another form factor (shape) is preferred. For contact cards, there are two smaller formats, known as ID-00 (minicard) and ID-000 (SIM card format) – see Figure 6.2.

The ID-00 ('minicard') size is not widely used in consumer applications, but it is sometimes issued as a 'companion card' alongside an ID-1 card, usually for marketing reasons. It appears to be gaining in popularity for contactless banking and loyalty cards in some market areas.

As SIM cards become embedded in more and more devices, particularly for data-only applications, even the ID-000 'postage stamp' size is too large to be ideal, and so within ETSI's Smart-Card Platform (SCP) project a new, 3rd form factor (3FF) was defined, known as the mini-UICC. The mini-UICC has half the area of an ID-000 card (a 'plug-in UICC' in ETSI terminology), but retains the same contact configuration and area: see Figure 6.3.

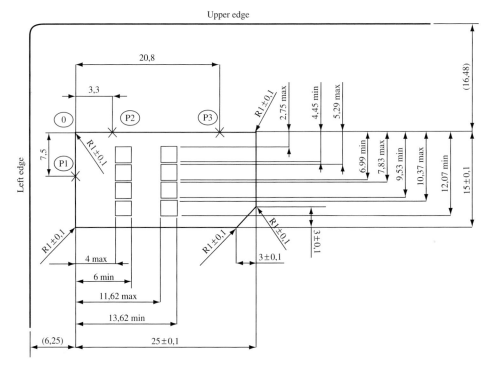

Figure 6.3 Mini-UICC format (© European Telecommunications Standards Institute 2006. Further use, modification or, redistribution is strictly prohibited. ETSI standards are available from http://pda.etsi.org/pda/ and http://www.etsi.org/services_products/freestandard/home.htm) (dimensions in mm)

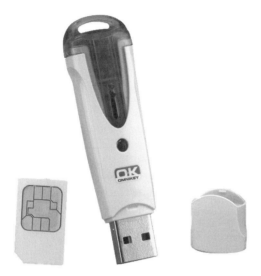

Figure 6.4 USB token (courtesy of Omnikey, GmbH)

For use with PCs, a combination of smart-card chip and reader, with a USB interface (see Figure 6.4), is often useful. This can perform authentication functions without the need for a dedicated smart-card reader.

For contactless cards, there are even fewer limitations, and almost any form factor can, in principle, be used. Smart-card chips have been built into mobile telephones, watches, 'buttons' and rings – see Figure 6.5. However, it may still be important to be able to test the device using some recognised and agreed test standard, and this becomes difficult if a wide range of devices is used.

Figure 6.5 Reverse (underside) of Paypass-enabled watch (courtesy of LAKS, GmbH)

Some card issuers – in particular where they are seeking a marketing 'angle' – also issue cards using other, non-standard, shapes but this does introduce a range of issues, for example:

- How do we ensure interoperability with devices built to the ISO standards?
- There is a range of tests for ISO 7810 cards, which ensure the durability of the card. If any characteristic of a card is changed, these tests can no longer be used and other characteristics of the card may be affected.

Smart-card chips can be built into other form factors – for example, a watch or the battery cover of a mobile telephone. While these represent very small volumes they are not considered part of the mainstream smart-card market, but they could be considered part of a wider 'portable identification' market. One form factor that is rapidly growing in importance is the passport, which I will discuss in more detail in Chapter 17.

6.3   Interfaces

6.3.1   USB

ISO 7816-12 defines a USB interface for a smart card, using two contacts that are left unused in ISO 7816-3. This may be used in a stand-alone device, as shown in Figure 6.4, or it can allow the card to be used in a conventional smart-card reader-writer, but with the USB contacts connecting to a USB interface device, to offer high-speed communications and a standard interface to PCs and similar devices. This could be important in a multi-application environment where secure authentication (using the ISO 7816-3 interface) is combined with a rôle for the card in an embedded system or PC network where larger quantities of data (such as a secure file or personal configuration setup) may be transferred under operating-system control.

6.3.2   Contactless cards

Contactless cards following ISO 14443 use an RF field at 13.56 MHz both to power the card and for communication between card and reader. Data transmission speeds are higher than for contact cards: the normal range is from 106 to 424 kbps (with 106 kbps being the default). This can be quite important for contactless card transactions, as the transaction must be completed quickly to avoid 'tearing' – where the card is removed from the field before the transaction is completed.

Cards are normally designed to operate at up to 10 cm from the reader antenna, although it is also possible to design card and antenna combinations that require much more specific placing of the card, and this may be preferred in some cases, to avoid the danger of accidental reads or collisions (where two or more cards are in the reader's field at the same time). However the ISO 14443 protocols themselves do include two different ways of handling collisions (corresponding to the two different protocols: type A and type B).

Type A is derived directly from the protocols used by the NXP MiFare^{TM} wired-logic card. The type B protocol was originally developed by Motorola but was adopted early on as an international standard, and has become the dominant standard for microprocessor-based contactless cards. MasterCard's 'Paypass' technology (which has also been licensed by Visa and other payment schemes) is based on ISO 14443 type B. Several other manufacturers sought, unsuccessfully, to have their proprietary protocols accepted as international standards, and these are still often used in proprietary products (and sometimes referred to, incorrectly, as ISO 14443 type X).

However, ISO 14443 allows the reader firmware to detect which protocol is being used by the card, and many readers can support not only type A and type B but also Sony FeliCa^{TM} and other proprietary protocols.

The simple operation and convenience offered by contactless technology conceals complex RF and signal-processing issues. Those who work in this field now find it relatively easy to establish basic communication with a card under most circumstances, however controlling and meeting exact parameters is considerably more tricky, and implementing a specification such as MasterCard's Paypass requires much detailed design and testing. Each time a small detail is changed (for example, the shape of the antenna) this has implications that go right back to the chip itself and its firmware.

The resolution of these issues is outside the scope of this book, but further details of contactless technology can be found in [3].

6.3.3   Dual-interface cards

For multi-application cards it is often necessary to support both contact and contactless interfaces. The individual applications can be set up to support only one interface, or both, while within an application some functions may still be reserved to one interface or the other (for example, loading a stored-value function may be possible only through the contact interface). In each case, the terminal should be aware of the rules, in order to display suitable user messages (otherwise it may just confuse the user with a 'read error' message).

It is a major advantage if the card can be completely personalised through one interface or the other. Two-stage personalisation raises issues of synchronisation; cards can be damaged during the second stage; and in some cases it is even necessary to introduce a third stage, to confirm that both the other stages have been completed. This significantly increases the length and complexity of the personalisation process.

6.3.4   Legacy contactless card emulation

Most first-generation contactless cards were wired-logic rather than microprocessor cards. In particular, the NXP MiFare^{TM} product gave 1 kB or 4 kB of memory, protected by sector keys and a proprietary algorithm.

As card and security technologies have evolved, it is becoming necessary to upgrade many MiFare^{TM} applications, and one option for this is to emulate the memory card functions. This requires a microprocessor card running the ISO 14443 type A protocol (the same low-level communications and anti-collision protocol as MiFare^{TM}), with a section of memory reserved for the card being emulated. When the card is powered up, it recognises through the ATR processing that the reader is attempting to read the memory and processes the commands accordingly.

Similar emulations can be designed for other memory and wired-logic cards.

6.4   References

[1] Riley, G. A. *Flip Chip Tutorials*. http://www.flipchips.com/tutorials.html
[2] Lau, J. H. (ed.) *Flip Chip Technologies*. McGraw-Hill 1995
[3] Finkenzeller, K. *RFID Handbook*, 2nd edn, Wiley 2003

7 Readers and terminals

The basic structure of a reader for contact cards was described in Chapter 3. This chapter considers the specific requirements of different sectors and of multi-application cards.

7.1 Reader type

In Chapter 3 we saw that readers may be manual or motorised, partial or full insertion, chip only or hybrid.

Motorised readers have specific advantages in multi-application environments too: the terminal can execute both 'warm' and 'cold' resets (see Chapter 8), allowing it to switch between applications without giving potentially confusing messages to the user or card-holder.

Many smart cards carry a magnetic stripe as well. This can be read when the card is inserted or when it is withdrawn; the former has advantages if some form of fallback is required, but reading a magnetic stripe on entry is often less smooth than reading on exit, and so gives slightly lower success rates. In retail environments where reliable reading of both chip and magnetic stripe cards is very important, special readers have been developed (see Figure 7.1) that combine a long reading slot for swiping with a 'park' position for reading the chip.

Contact readers must also have limit switches or other methods for detecting when a card is in place; these are used not only for powering up the card but also for detecting when a card has been inserted wrongly or not removed at the end of the transaction; in these cases it is often desirable for the terminal to emit a warning tone or signal.

If a card does not power up correctly, this can sometimes be attributed to poor electrical contact or card positioning; in these cases an insertion reader can ask the user to re-insert the card, or a motorised reader can reverse and try again.

7.1.1 Antennae for contactless readers

The 'reader' for a contactless card is called a proximity coupling device (PCD) by ISO 14443.

Figure 7.1 'Swipe and park' reader (courtesy of VeriFone Inc.)

Most antennae used in contactless readers are of the planar type – they consist of specially shaped tracks on a printed circuit board. They are linked to logic and signal processing to detect and manage cards in the field, and to handle the communications protocol (known as T = CL) and, in particular, the anti-collision scheme.

ISO 14443 readers are normally multi-protocol; that is, they can handle both type A and type B cards (and often other proprietary protocols such as Sony FeliCa™ as well). Cards should normally remain silent until polled, and this means that it is possible to have several type A and type B cards in the field at the same time, with only one card communicating with the reader. The ISO 14443 protocol defines an envelope that allows normal ISO 7816 packets (known as application protocol data units or APDUs) to be exchanged between card and reader; the rest of the application is therefore unaware whether it is dealing with a contact or a contact-less card.

7.2   Terminals

In most cases, the smart-card reader is integrated into a terminal device; these cover a very wide range of functions and physical forms, and this is only a small selection of terminal types:

7.2.1   Point of sale

Smart cards are used at the point of sale (POS) for payment, to track customer loyalty and staff, and for loading value into other systems.

Point-of-sale equipment ranges from simple desktop cash registers to supermarket checkouts that must also integrate scanning, weighing and stock-management functions. There are variations for special environments such as restaurants, bars, hotels and petrol stations. And since business practices and laws vary from country to country, often requiring a special design, the range of models is immense.

Most modern electronic point-of-sale (EPOS) terminals incorporate powerful PCs that can handle both attached devices and networked communications with intelligent subsystems. Electronic cash registers (ECRs) have less functionality but commonly incorporate a simple protocol for connecting to a payment terminal; this allows the till to send a total for payment to the terminal and to receive confirmation that the amount has been paid. In the first case, the smart-card reader may have a few extra functions, as the application software runs on the EPOS terminal. In the second case, the smart-card reading device must have enough processing power to run the application and communications devices to access databases and host systems.

Even where the POS terminal is PC based, there may be advantages in using an intelligent smart-card reader-terminal. If the volume of data to be exchanged is large, then performance may be better in a dedicated unit. Or when data or keys must be held within a secure unit, it is easier to demonstrate the security of a smaller device. The smart-card terminal may incorporate security mechanisms that protect it from most attacks, including having rogue software downloaded to it.

In a multi-application environment, this protection from attack will be very important: the card issuer must trust the application that will read its cards. However, where the retailer is itself an issuer, for example for its loyalty cards and staff cards, the ability to change software freely and to interact with other applications on the POS terminal may argue for holding applications on that terminal. Most advanced retail applications now use a hybrid architecture, with keys and certified applications residing on the smart-card terminal and other applications on the point-of-sale system.

7.2.2   Vending

Vending machines are dedicated, self-service point-of-sale terminals. Smart cards are most likely to be used for payment, and may be either 'open' payment cards (such as Visa or MasterCard) or campus cards used in a business, school or other organisation.

These each pose quite different design and security problems: in the first case the card and acceptance system are defined in great detail; they require card-scheme keys and encryption systems, and the device must be approved by the card scheme (see Chapter 13). A bought-in original equipment manufacturer (OEM) device is likely to be the most cost-effective way of achieving this. Several companies make units that can be built into a vending machine, replacing or alongside the coin slot mechanism – see Figure 7.2.

Figure 7.2 OEM card accepting device (courtesy of VeriFone Inc.)

For campus cards, the requirements are simpler and most of the processing is usually performed on a host system; this requires good online communication and online security.

7.2.3   Kiosks

Kiosks are self-service devices that are often specifically intended to cater for a wide variety of functions and applications. They are normally PC based and card reading is a standard function. Again, the issue with these devices in a multi-application environment is the need for multiple certifications: OEM devices are likely to be most suitable but some suppliers have made a niche business from creating multi-application kiosks to meet the needs of a range of contact and contactless applications.

7.2.4   PC-connected readers

For some domestic, industrial and commercial applications the only economic way to support smart-card applications is to use a smart-card reader directly connected to a PC as a peripheral, via a serial port, USB (universal serial bus) port or PCMCIA (PC card) interface. The standard for this is known as PC/SC and is managed by an industry body: the PC/SC Workgroup [1].

PC/SC specifications (by permission of the PC/SC workgroup)

The PC/SC specifications 2.01.2 are divided into ten parts (Figure 7.3). A brief summary of each part is provided below:
- Part 1: provides an overview of the system architecture and components defined by the workgroup;
- Part 2: details compliant ICC-IFD (smart card – interface device) characteristics and interoperability requirements;
- Part 3: describes the interface to, and required functionality for, compliant IFD; there is also a supplement to provide information on RID numbers;
- Part 4: discusses design considerations for IFD devices; in particular, it provides a recommended implementation for PS/2 keyboard-integrated IFDs;
- Part 5: describes the interfaces and functionality supported by the ICC resource manager, a required system level component;
- Part 6: describes the ICC service provider model, identifies required interfaces, and indicates how this may be extended to meet application domain-specific requirements;
- Part 7: describes design considerations for application developers, and how to make use of the other components;
- Part 8: describes recommended functionality for ICCs intended to support general-purpose cryptographic and storage requirements. This is oriented towards support of internet and PC standards for security and privacy;
- Part 9: describes the management of IFDs with some extended capabilities such as secure PIN entry or user interface functionality;
- Part 10: describes the management of IFDs with secure PIN-entry capabilities.

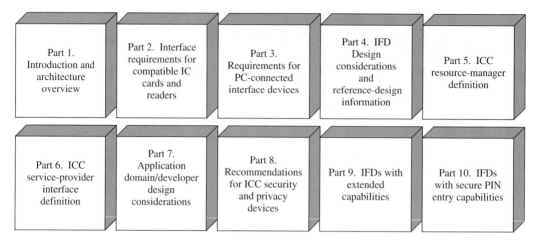

Figure 7.3 PC/SC specifications (courtesy of PC/SC Workgroup)

The PC/SC specifications are continuously being updated; version 2.01.3 was issued in January 2006. The specifications build on the ISO 7816 and EMV (card-payment industry) standards; they are quite detailed and include formal definitions of logical devices, services and classes – see box.

Since the addition of cryptography and PIN entry features to the PC/SC specifications, devices built to these specifications can meet most of the requirements for technical interoperability in typical commercial applications.

However, a PC is not generally considered a secure computing environment (particularly if it has an internet connection) and so many applications will have additional requirements for security. And as we will see in Chapters 13–18, many application groups have their own certification and type-approval procedures for terminals.

7.2.5 Access control

Access control terminals are amongst the simplest smart-card devices. In addition to the card reader, they are likely to include a keypad or biometric reader and, of course, the relay or interface to open the door or gate. They will also have a connection to a host system, although usually systems are configured in such a way that in normal use the identity verification and access-permission checks are carried out locally, for reasons of both performance and reliability.

Most access control systems are proprietary and there are no certification requirements.

7.2.6 Personal smart-card readers

For some applications, there is a need for offline readers, which can, for example, read a balance or generate a security code. There is growing demand for these devices, for use in internet security applications such as e-banking logon.

Figure 7.4 Balance reader (© www.smartcardfocus.com)

The very smallest devices (see Figure 7.4) need only cover one corner or edge of the card; the balance or other information is displayed on a small LCD (liquid-crystal display) screen. Other readers (as shown in Figure 7.5) have a small keypad and display; this allows the user to select functions or enter a PIN.

7.3   Terminal management

Card-reading terminals have become considerably more complex over the years. In the 1990s they moved from fixed logic to a software or firmware basis; nonetheless, the software itself was relatively simple (usually less than 64 kB of memory were required) and there were few variable parameters.

With the advent of chip cards, there has been a further step in complexity. Terminals that support these cards have anything from 256 kB up to 8 MB of memory; most support multiple application programs and for many applications they must include cryptographic functions to secure the terminal itself and the data being exchanged between the terminal and host computer.

The terminals themselves conceal this complexity by using software that makes the new functions transparent to the user, but there is now a need to manage the additional software and parameters.

For magnetic stripe terminals, the manufacturer almost always produced the software. It was probably written in a low-level language and consisted mostly of drivers for the various devices on the terminal and a fairly simple logic flow; as terminals were sold into new markets, the prompts could easily be adapted for different languages and retail environments. Chip-card terminal software is much more specialised and each new card application requires new terminal software.

There is, therefore, a move towards using specialist software houses to create the terminal-application software; these prefer to work in high-level languages so that they can sell their applications to many customers, including other terminal vendors. This, in turn, leads to a demand for standardised terminal structures and interfaces, and led to the creation of the Small Terminal Interoperability Platform (STIP), which is described in more detail in Chapter 9.

Figure 7.5 Personal card reader (© Vasco 2006)

With the older terminals, software changes were rare and there were no crypto-graphic keys. It was, therefore, feasible to maintain these terminals using manual methods (engineer visits or postal exchange).

Software for chip-card terminals, however, evolves continuously, as new products are developed and new features introduced on current products. The more complex software may have bugs, or it may be found that users prefer slightly different wording on a prompt. Terminal owners should expect at least one software update every six months.

Where cryptography is used to secure data or transactions, the terminals must store keys for each application or card issuer, often in several lengths and versions. If a key expires or may have been compromised, then it must quickly be removed from terminals and replaced with an updated key. Because of the limited time available to

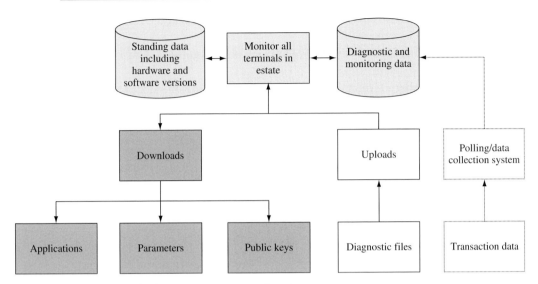

Figure 7.6 Terminal management system functions

make these changes, and the need to provide an audit trail and verification process, it is essential that they are made electronically.

These new terminals are powerful computers which, if only performing one application, sit idle most of the time. Multi-application schemes allow the terminal to perform many more functions, and often improve the economics of terminal placement. The terminals must be programmed to make use of these extra functions; new functions and ways to add value may appear quite frequently, and to take advantage of these opportunities it is important that whole terminal estates can be upgraded quickly.

Against this background, owners and operators of even medium-sized terminal estates (from a few hundred terminals upwards) have found that the old, manual methods of updating terminals are inadequate; a terminal management system (TMS) is required.

The main functions of a TMS (see Figure 7.6) are to:

- Keep track of all the terminals in a terminal owner's estate, including the hardware and software versions, public-key versions and facilities enabled;
- Manage the efficient download of applications, parameters and keys to all terminals;
- Once a new application, application upgrade or parameter change has been downloaded, trigger a changeover to the new software or advise the user to perform an action to effect the change;
- In the case of large downloads (e.g., a whole application), to compress or divide the download to minimise the time required, and to handle any errors arising during the download;

- Respond to terminal requests for downloads (e.g., at initial installation or following a fault);
- Periodically upload (collect) monitoring and diagnostic files from terminals where the terminal generates these.

Many modern terminals are programmed in platform-independent languages such as Java and C++. With these terminals, it is increasingly possible to develop and run applications that will run on several different terminal types. Some of these applications may be for specialised functions or specific services, and may be developed by software companies other than the terminal supplier. In these cases the TMS must also handle application certificates and the application uploading process.

7.4   Reference

[1] PC/SC Workgroup. www.pcscworkgroup.com

8 Application selection: the ISO 7816 family

We saw in the previous chapter that the application selection function plays a pivotal rôle in the design of any card-reading terminal. In this chapter, I will review the requirements for application selection and the options for cards and terminals to implement this function.

8.1 Scope and functions

Application selection is required for any microprocessor card, not only for multi-application cards, since it is the process by which the card application is started up.

For memory and wired-logic cards, even though there is no firmware 'application' on the card, the initial process is similar but application selection is implicit. In this case the terminal must select the appropriate supporting program and functions.

Where either the card or the terminal is multi-application, in the widest sense described in Chapter 2, application selection plays a pivotal rôle in determining which of the applications will be run; it links the technical protocol-handling functions with the logical transaction flow.

In some cases, the application to be selected is known before the card is inserted, either from the context of the transaction, or because an operator or user has already selected a function. In other cases the card and terminal must agree the selection, or offer a choice to the operator.

8.2 Card initialisation

8.2.1 Power up and reset

The process starts with powering up of the card and the way the card responds to the reset signal: its answer-to-reset (ATR).

ISO 7816-3 defines the sequence (see Figure 8.1), starting with an unpowered card and all the terminal outputs, either unpowered or in their LOW state:

- The RST circuit is put into the LOW state;
- Power is applied to VCC. If the terminal supports 3 volt operation, it will start with 3V; otherwise it will directly apply 5 V power;

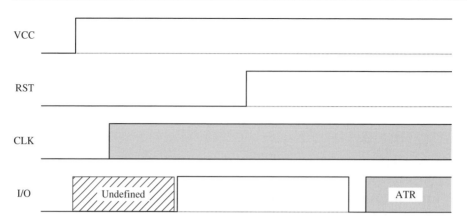

Figure 8.1 Power up and reset sequence

- The I/O circuit is put into reception mode;
- A clock signal (1 to 5 MHz) is applied and the terminal waits at least 400 clock cycles for it to stabilise;
- The RST circuit is put into the HIGH state.

There are two types of reset: a cold reset, when the card is first powered up, and a warm reset, which may take place at any other time. Warm resets allow the terminal to move between applications on the card; they are, therefore, very important in a multi-application world, but are often not implemented correctly. When a warm reset takes place, any protocol, speed or voltage negotiated with the card during an earlier application remains in use, but other parameters (such as a file selection) are reset; there are sometimes issues where either the card or the terminal resets either too few or too many parameters during a warm reset.

The card should respond to the reset with an ATR: a string of up to 33 characters that declares its capability and preferred operating mode to the terminal. (If the card is only 5V capable then it will wait until it is powered at that voltage before sending an ATR.)

8.2.2   Content of ATR

The answer-to-reset consists of two mandatory initial characters, followed by up to 31 further characters – see Figure 8.2.

The first character (TS) sets up the bit speed and coding convention (direct or inverse) being used by the card; this allows the terminal to parse all the subsequent bits (or 'moments') into bytes (so from now on we can call the characters bytes).

The next byte (T0) tells the terminal whether there are interface bytes or historical bytes, or both, in the ATR. For most multi-application cards there will be some of each.

The *interface bytes* set up the transport layer of the card-to-terminal interface; they allow the card to request specific values to be used on this interface, for example a

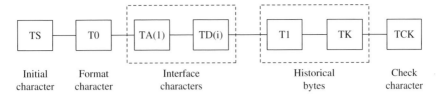

Figure 8.2 Structure of answer-to-reset

clock-rate or baud-rate adjustment factor, extra guard time, a programming voltage or maximum current (these are rarely used). Most important, they determine whether the card uses the byte-oriented protocol, known as $T = 0$, or the block-transmission, $T = 1$, protocol.

The interface may operate in either the 'specific' mode (where clock speed and voltage are explicitly set to the values requested by the card) or 'negotiable mode' (where the terminal may offer another set of values). If the terminal wants to change these modes or values, then it must issue a 'warm reset' to the card.

Once the transport layer has been set up, the *historical bytes* tell the terminal how to work with the card. There may be up to 15 bytes of historical data, including:

- A country code (more often this is coded in the application data);
- The issuer identification number (of which we will hear more later);
- Card service data: this data object indicates by a bit map which application selection methods the card supports: direct, partial or implicit (see next section);
- Card capabilities: this bit map indicates what types of file access the card supports (direct, partial, implicit, file or record number);
- Initial access data: this object allows the card to send an 'initial data string' to the terminal (note that this is before an application has been selected and so should apply to all applications on the card; it may indicate a card state or status. The initial access data, however, may also be provided at other times);
- Card issuer's data: this is a free-format field controlled by the card issuer;
- Pre-issuing data: this data object may be used to specify the card manufacturer, IC type, ROM or operating-system version.

In principle, all these objects are optional; however, several of them will, in practice, be present for most multi-application cards.

8.2.3 Protocol negotiation

In its ATR, the card may either offer the *specific mode* (and with it a request for a particular clock frequency, baud rate and transmission protocol) or else allow negotiation. If the terminal cannot support the specific parameters offered, it must perform a warm reset, in response to which the card will normally enter negotiable mode.

In negotiable mode, the terminal uses the protocol and parameter selection (PPS) protocol to offer a change of protocol to the card; the card normally accepts by echoing the request. Once accepted, this protocol and these parameters are used for all further communications until the next reset.

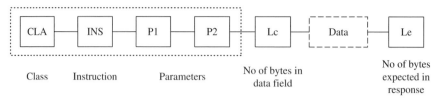

Figure **8.3** Structure of command APDU

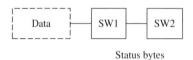

Status bytes

Figure **8.4** Structure of response APDU

8.2.4   Commands

Commands and responses are exchanged between terminal and card using a structure known as an application protocol data unit (APDU), defined in ISO 7816-4. There are two types of APDU: command and response APDUs. Command APDUs start with a four-byte header (CLA INS P1 P2), where:

- CLA is the class of command (7816-4 or other, and whether secure messaging is used);
- INS is a one-byte instruction code: ISO 7816-4 defines 18 instructions, but other instructions may be used where the CLA byte specifies a proprietary command;
- P1 and P2 are two parameters; if not required by the INS code, they are set to 00.

The header is followed by a length field and any data sent as part of the command, and finally a length field showing how many bytes of data are expected in the response (see Figure 8.3).

Response APDUs consist of the requisite number of bytes of data, followed by two status bytes: SW1 and SW2 (see Figure 8.4). If there are no data, then the response consists of two bytes only.

If an application uses only ISO 7816-4 instructions, then, in principle, any microprocessor card can process that application, so no special software is required on the card. This allows applications that only need to read from and write to the card memory to be constructed very simply. In practice, however, each sector defines its own instruction classes and instruction sets, through its own standards (e.g., ETSI for telecoms, Calypso or ITSO for transportation, EMV for banking). Even in the case of national ID cards, many implementations deliberately make use of proprietary instructions as a part of the security scheme.

8.2.5   File selection

Files in a smart card are arranged in a hierarchy, which is also an important part of the security structure. The highest level is the master file, of which there is only one on

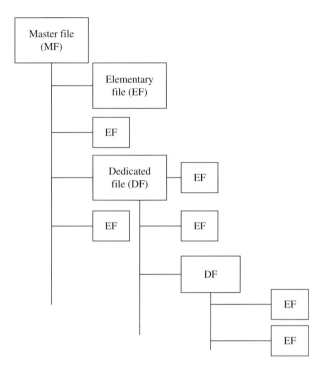

Figure 8.5 Smart-card file structure

each card; below this may be several layers of dedicated files (DF), and finally one layer of elementary files (EF) – see Figure 8.5. Access to the lower levels must pass through a logical channel comprising each of the parent files in turn. Each layer contains the access rules and control information relating to the next layer down, and each file has associated with it a secure key-space that maintains the authentication and confidentiality of stored data.

A key INS code is the 'select file' command (A4). This allows the terminal to set up a logical channel directly to a dedicated or elementary file, with the appropriate level of security. For 'inter-sector' applications this is the way data may be shared or passed between applications, and for 'generic' ISO 7816-4 applications it is the only form of application selection required.

8.2.6   Further application selection requirements

Many proprietary or sector-specific applications, however, have their own special requirements for application selection. For example, EMV payment cards make use of a further level of application selection once the payment systems environment DF has been selected. The same is true of mobile telephones making use of the SIM toolkit (described in Chapter 14) and the ITSO interoperable public-transport structure.

As with the proprietary commands described in the previous section, these represent an extra layer of complexity for the multi-application card designer, but by no means an insurmountable barrier. What is usually needed is a master application, linked to the master file, that is aware of these applications and their special requirements.

This application is usually called a card operating system, although its functions are much less comprehensive than those of a normal computer operating system since there are simply many fewer resources to control.

8.3   Native operating systems

Most current cards use an operating system designed by the card manufacturer, which directly calls functions in the hardware; this is called a *native* operating system.

Native operating systems are usually optimised for that particular card or card family, and they can be resident in the 'hard mask' of the chip – the functions that are built in to the card at the time of manufacture – or in PROM. They are therefore relatively efficient both in memory usage and in performance. However, any application must be written to take account of the operating-system features, and often in a low-level language that is specific to that card manufacturer or operating system.

These operating systems are, therefore, most suitable for managing relatively stable functions that are always present in this card type (for example, a card that has been optimised for access control). From the card manufacturer's point of view, a native operating system is under its direct control; it can make changes and introduce new features whenever required. This helps to provide some differentiation between card manufacturers, and thereby avoids competing only on the basis of price. The native operating system provides the interface with the specific features of the card's hardware, and calls any special functions supported by the card.

From the card issuer's point of view, however, there are several drawbacks to using native operating systems:

- They tie the application to a particular operating system and, hence, often to a single manufacturer. Although most card manufacturers do have cross-licensing arrangements with other suppliers, invoking those licences does carry an extra cost, and so this is usually only done where there are production problems;
- The security of the application is in the hands of the operating system, and hence those of the card manufacturer. Although several native operating systems do have evaluation certificates under Common Criteria or an equivalent structure, many card issuers want to integrate and control the security of their own cards into their overall security structures, for example by managing the keys that allow applications and data to be downloaded into cards;
- The software development tools available under native operating systems are usually quite primitive compared with those available for mainstream languages, and the number of programmers who can write card applications is much smaller: this reduces competition and increases development timescales;
- All the applications that will share a card, and in particular those that will share data, must trust one another or be exhaustively tested together: this also slows innovation and increases time-to-market for upgrades and new products;

- Where the card issuer wants to be able to make changes to an application, these must be sent as software 'patches' to the card and stored in the precious non-volatile memory (usually E^2PROM). This works reasonably well for small numbers of such patches, but if regular changes or updates are made, then a mechanism for downloading complete applications or sections of an application is very desirable.

In the next few chapters we will explore how a group of specifically multi-application operating systems overcomes these difficulties. However, each of these operating systems does, to a greater or lesser extent, increase the cost of the card, either directly through licensing fees, or through the higher overheads and memory requirements they bring. Users must, therefore, balance the cost of the card against the advantages of using a multi-application operating system.

9 JavaCard and GlobalPlatform

9.1 History

In 1996, smart-card manufacturer Schlumberger demonstrated a card operating system to which it had added Java bytecode interpreter functions for a small number of methods. This initial implementation was very limited and involved an intermediate format, but it attracted considerable interest because it offered, for the first time, an opportunity for mainstream computer programmers to become involved in smart-card application development.

At the same time as Schlumberger was working on this development, Visa was working with Integrity Arts (a subsidiary of another smart-card manufacturer, Gemplus) to specify an 'open' smart-card operating system that could work on any manufacturer's card, permit the use of programmer-friendly development tools and allow applications to be downloaded to the card.

The two streams of activity together triggered Sun Microsystems to buy Integrity Arts and to endorse a specification for a Java implementation on a smart card, known as JavaCard 1.0, which drew on both efforts. Gemplus and Schlumberger joined with other smart-card companies to form the JavaCard Forum, which released the JavaCard 2.0 API at the end of 1997. This second release was considerably more detailed and allowed many more practical implementations.

However, even this version did not ensure portability of applications between smart-card platforms, and did not define in any detail the mechanisms for downloading applications to the card. To overcome these limitations, Visa published its Visa Open Platform specification in 1998, which defined mechanisms for secure applet download and on-card management. In 1999, Visa ceded control of the Open Platform specifications to the newly formed GlobalPlatform organisation [1], which includes telecoms companies as well as payment schemes, smart-card vendors and integrators.

GlobalPlatform has, since that time, considerably expanded the scope of its activities; instead of focusing entirely on the card itself, it now addresses interoperability standards for cards, devices and systems. Although most GlobalPlatform cards are still based on the JavaCard structures, there is also an API published in C for Multos cards, while the systems specifications that cover issuance, downloading and personalisation can be – and are – used with Multos or native operating-system cards.

Sun continues to be responsible for the JavaCard specifications [2], although the JavaCard Forum [3] has an advisory rôle and also publishes associated specifications such as a biometric API for JavaCards. A separate forum does continue to exist for OpenCard (general-purpose APIs built on JavaCard), but the main focus of activity depends on the specifications now being issued and maintained by GlobalPlatform.

9.2   JavaCard

9.2.1   Scope and components

Java is an interpreted language that has been designed to meet the criterion 'write once, use anywhere'. JavaCard performs the same function, but within the limited environment of the smart card: Java programs written for a browser environment will not generally run in a smart card, because the card has much more limited resources (for example, it has no display). Nor are JavaCard applets upwards compatible; they have been designed to meet ISO 7816 file naming, application selection and file structures, and they take account of the specific operational and technical features of the smart card: for example, by distinguishing between RAM and E^2PROM for data storage.

JavaCard defines (see [4]–[8]):

- A subset of the Java language that removes those functions not relevant to smart cards and other small embedded systems;
- A structure for defining the runtime requirements of the program, for compiling and testing it and generating the Java bytecode (the 'compiled' or pre-processed Java program);
- A runtime environment that enables the Java bytecode to run on each manufacturer's hardware.

The last of these is the responsibility of each smart-card manufacturer that offers a JavaCard interface, as it depends on the particular features offered by the hardware and card operating system components – see Figure 9.1.

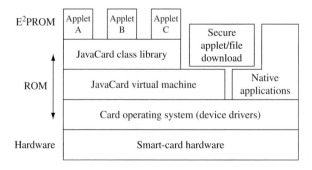

Figure 9.1 JavaCard structure

9.2.2   Applets

A JavaCard application consists of an applet (the bytecode stored on the card), which calls a set of class libraries. A JavaCard applet is not as dynamic as a Java applet – it cannot be loaded and used in real time – but it is given the name because it can be downloaded to the card after the card has been issued, and is likely to be stored in dynamic memory (usually E^2PROM in today's smart cards).

The system class libraries are normally stored in ROM, as a given card will conform to a particular issue of the specification; other (user-defined) classes are usually loaded into E^2PROM. JavaCard system class libraries cover the range of functions needed by JavaCard programs; this is a much smaller range than for Java applets because of the smaller range of classes and methods allowed by JavaCard. For example, JavaCard does not use floating-point data types, strings or multi-dimensional arrays, and threading is not supported.

This structure offers two of the key benefits of JavaCard: *portability* and *familiarity*. In principle (and increasingly in practice), JavaCard applications are portable across all JavaCard implementations from different manufacturers. And many programmers are familiar with Java structures and language from other applications, making it easier to move into smart-card programming from other parts of IT.

9.2.3   JavaCard Virtual Machine

The Java language makes use of a *virtual machine* concept; a Java Virtual Machine (JVM) has all the knowledge and resources it needs to run Java bytecode in a particular hardware environment. The JavaCard virtual machine (JCVM) differs from most JVMs in that it is divided into two: the smart card only contains those functions relevant to the runtime execution, while other functions such as class loading, linking and bytecode checking are carried out by a *converter* program, running on a PC platform or workstation.

This structure allows the JCVM to be as small as possible – another key requirement for a smart-card environment.

The development toolkit usually also contains a JCVM emulator, which allows the programmer to test most of the application functions (but not, in most cases, the interaction between applications) on the PC platform before loading it in to the card. It also contains the host end of the installation package needed to load applets to the card. Figure 9.2 shows a typical JavaCard development environment.

9.2.4   JavaCard runtime environment

The JavaCard runtime environment (JCRE) is the part of the JavaCard structure that must be provided by the smart-card vendor: the hardware, card operating system, runtime components of the JCVM, and the class libraries. There exist reference

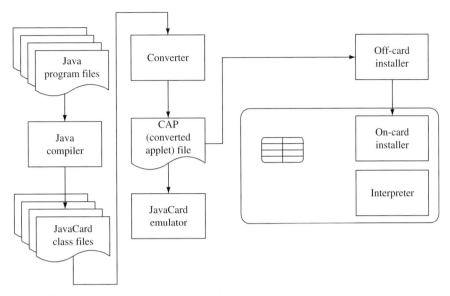

Figure 9.2 JavaCard development environment

implementations of the last two, and it is now a routine process for a smart-card vendor to produce its own JCRE.

The JCRE also usually contains the on-card half of the installation program; as we have seen, JavaCard itself does not fully define this process, and this is where GlobalPlatform comes into play. You can have a JCRE that does not contain an installer; in this case the applications must be loaded during the card initialisation or personalisation process. These cards are known as JavaCard-S, or static JavaCard; they are generally cheaper than other JavaCards and may be a good choice in those operating environments where post-issuance download is not an issue; we will see in Chapter 13 that this is quite common.

An important feature of the JCRE is that it runs all the time that the card is powered (unlike a conventional JVM). Each time the card is powered up, the JCRE restarts and retrieves its last state from memory, which allows the card to recover from any data errors or 'tearing' (where the card is removed from the reader before the transaction is complete).

9.2.5   Security model

A highlight of JavaCard is its security model, which gives the card issuer a high degree of flexibility as to the level of control it exercises over applets and data objects on the card. The bytecode checker, which forms part of the converter, seeks to ensure that no application uses resources outside its range. The converted applets (known as the CAP file) are signed as a package by the issuer using a DES secret key, and the card checks this signature (using the same key) when the CAP file is loaded.

For runtime checking, instead of the *sandbox* normally associated with Java, JavaCard uses a software firewall mechanism that explicitly links each object with the applet that owns it, and prevents access to those objects by other applets. These two mechanisms together allow the card issuer to design security structures that allow multi-application cards with one or more application issuers and loading channels, and with levels of security appropriate to the application.

These features are defined in more detail in the GlobalPlatform specifications; although a static JavaCard (JavaCard-S) can be implemented without GlobalPlatform, a much wider functionality is obtained by implementing JavaCard and GlobalPlatform together.

9.3   GlobalPlatform

9.3.1   GlobalPlatform architecture

GlobalPlatform is a consortium that brings together smart-card manufacturers, software companies, and user sectors (mainly finance and telecommunications). It has a broad remit, describing itself as 'the standard for smart-card infrastructure' with a mission to 'establish, maintain and drive adoption of standards to enable an open and interoperable infrastructure for smart cards, devices and systems that simplifies and accelerates development, deployment and management of applications across industries' [9].

GlobalPlatform complements the JavaCard concepts by specifying the processes for loading and managing applications on the card; it also offers a standard terminal architecture offering similar benefits for multi-application terminals, and a standard command set for personalising cards (loading applications and parameters before the card is issued). Today GlobalPlatform can be implemented using any runtime environment that offers platform independence and on which the GlobalPlatform Environment can be implemented.

Both card and terminal may have multiple applications, but they need not always be the same groupings: see Figure 9.3.

The key to the GlobalPlatform architecture is its recognition of the rôles played by the different parties in a multi-application environment. Whereas for a single-application card it may, in principle, be acceptable for the card issuer to assume responsibility for all aspects of the card's development, production and use, for a multi-application card there is a complex set of relationships between the developers, implementors and operational users of the card. In fact, even for a single-user card there are significant technical advantages in, for example, ensuring that the issuer's keys are only used within the issuer's own domain and, contractually, it is important to be able to define responsibilities for a service level agreement.

Table 9.1 shows the rôles defined by GlobalPlatform, while Figure 9.4 shows an example of the way they might interact in a real example. Here, a bank and a public-transport operator have come together to issue a card that can be used to pay for both travel and other goods. New cards are issued with both the banking and transit applications, while the transit application can also be loaded to a bank card after

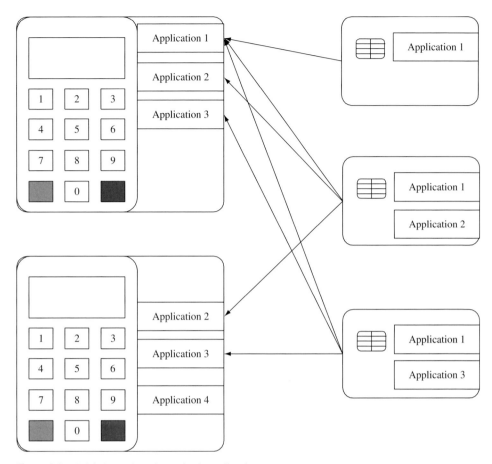

Figure 9.3 Multiple card and terminal applications

issuance. The bank's smart-card supplier also provides the GlobalPlatform environment and the payment application software.

9.3.2   Card Manager

This structure is enforced by the Card Manager, an application on the card that acts as the issuer's agent in controlling what applications may be loaded, like a gatekeeper at a remote site who only admits people on instructions signed by the Security Manager at Head Office, and who checks which areas they have been allowed to visit.

The Card Manager grants access to the card for sensitive functions such as loading a new or updated application, and checks that applications only use the memory allocated to them; it checks each command (APDU) arriving at the card and sends it to the currently selected application. Card Manager keeps track of the card's life-cycle and that of the applications, as described below, and manages the security domain structure. It also provides a Global PIN that can be shared between applications.

Table 9.1. *Rôles in GlobalPlatform*

Rôle	Main function	Typical actor
Card issuer	Issues card to card-holder; takes responsibility for the rest of the structure	Bank, transport operator, telco, employer or government
Card enabler	Initial card personalisation and key loading	Bureau or card manufacturer
Card manufacturer	Fabricates and tests card	Specialist manufacturer
IC manufacturer	Fabricates chips for use in smart cards	Semiconductor manufacturer
Platform developer	Creates card operating system	Smart-card manufacturer or specialist software company
Platform specification owner	Specifies card operating system	GlobalPlatform consortium
Application provider	Provides a service using a smart-card application	As for card issuer (but need not be the same company)
Application owner	Owns and licences IPR and AIDs for the application	Card scheme or consortium
Application developer	Creates application software	Specialist software company
Collator/decollator	Collates personalisation data for multiple applications and sends them to loader	Personalisation bureau or card issuer
Card-holder	Uses card, according to terms and conditions	
Loader	Loads applications onto card, either at initial personalisation or subsequently	Bureau or network operator

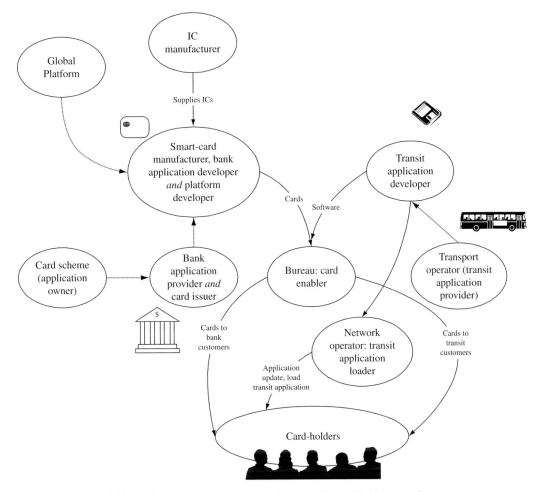

Figure 9.4 Example interaction between actors in a GlobalPlatform environment

9.3.3   GlobalPlatform API

The Card Manager application is complemented by an API on the card that is
embedded into the JavaCard Virtual Machine; it can also be embedded into another
card operating system, including Multos or .NET, thus allowing cards using those
operating systems to take advantage of the GlobalPlatform card and application
management structures – see Figure 9.5.

This API provides a small number of additional functions that are needed by
GlobalPlatform applications and by the Card Manager. Using these functions an
application may, for example:

- Lock the whole card if it feels that card security may be breached;
- Open a secure channel for off-card communications;
- Verify a key check value before loading the key to the card.

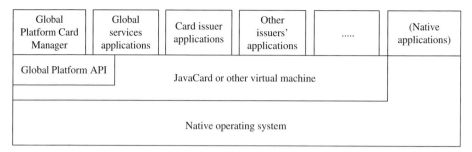

Figure 9.5 GlobalPlatform card architecture

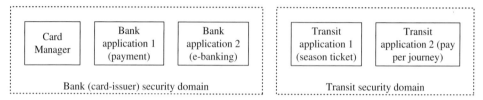

Figure 9.6 GlobalPlatform security domains

9.3.4   Security domains

The Card Manager controls the issuer's *security domain* on the card. It manages the issuer's keys, sets up a secure channel to the issuer where necessary and controls permissions for activities such as application loading.

GlobalPlatform also allows the issuer to set up additional security domains on the card. This is likely to be needed where one or more applications on the card are controlled by a completely different organisation: the public-transport application on a bank-issued card, or the open payment application on a SIM card. The card issuer must delegate specific responsibilities to the secondary domain, and the Card Manager will always check that the domain has the appropriate rights before allowing a sensitive operation, but if correctly implemented this removes the need for the card issuer to manage the other organisation's applications. From Version 2.2 of the GlobalPlatform card specification, these rights can be changed dynamically by the card issuer.

Each application belongs to a security domain, as in Figure 9.6, which represents the domains for the scheme described in Figure 9.4.

As we will see in Part III, mastering this separation of security domains, and the split of responsibilities that it entails, is a critical factor in managing multi-application cards.

9.3.5   Card life-cycle management

Another aspect of card and application management addressed by GlobalPlatform is the card and application life-cycle. It recognises that cards go through a sequence of states, as shown in Figure 9.7:

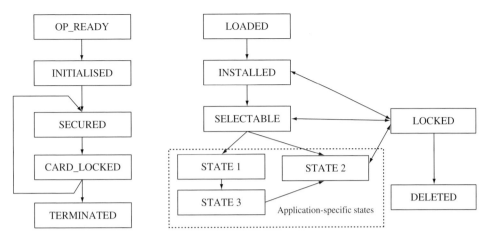

Figure 9.7 Card and application life-cycle

The card life-cycle starts when the GlobalPlatform API and Card Manager are loaded; it is next initialised by having further applications and parameters loaded to it, and is then secured prior to issue. If a security event is detected, the card may be locked, but can be unlocked (returned to the SECURED state) by the card issuer. If the card is terminated, then it can no longer be restored or used.

The life-cycle of an application is more specific to that application, and so is managed by the application itself using the functions in the GlobalPlatform API. Once the application is loaded into the card in the CAP format, it must be converted into the runtime format ('installed') and made selectable. At that point, it can be selected by a terminal application, until locked by the application itself or its owner.

Again, the correct management of card and application life-cycles is very important in a multi-application world: a bank's criteria for blocking a card may be very different from those of a bus company or a local government.

9.3.6   GlobalPlatform device specification

As we showed in Figure 9.3, it is possible to have multi-application cards with single-application terminals and vice versa. However, the GlobalPlatform vision is of a completely interoperable world in which terminal owners use similar principles for managing their terminals as card issuers use for managing their cards.

In practice, for many terminal types it is much easier to manage terminal applications, and indeed many of the systems in which smart cards are used already have a wide range of applications that must be managed: a point-of-sale system must have product line and price files, programs for handling special offers, new forms of payment or new legal requirements. A ticket-selling machine must have the latest fares and network data, hot lists and action lists, while a mobile telephone has many functions that do not call the SIM card.

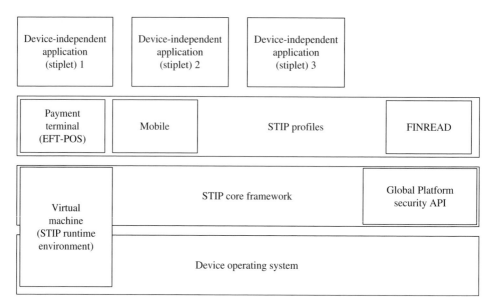

Figure 9.8 Small Terminal Interoperability Platform (STIP) device structure

However, there are some environments in which the terminal and its applications can be managed in a very similar way to a card, and it is certainly important to include the terminal in the security architecture when downloading applications or parameters to the card. We are most often talking here about stand-alone smart-card-reading terminals, although kiosks and similar devices may sometimes fall into this category.

The GlobalPlatform Device (GPD) specification is based on the Small Terminal Interoperability Platform (STIP) specifications, which we first met in Chapter 7; GlobalPlatform acquired the assets of the STIP consortium in May 2004 and now maintains the specifications. The objectives of this specification are to:

- Allow portability of code between terminal types: as with the JavaCard specification, an application developed for one GPD terminal should run without modification on another GPD terminal;
- Allow card and terminal applications to be developed and tested together;
- Reduce the cost and time required to develop terminal applications.

A STIP device (see Figure 9.8) contains:

- Its own operating system: this may be a proprietary operating system but will include a range of additional functions and device drivers for the device's peripherals;
- The STIP runtime environment: this forms the interface between the operating system and the STIP core framework;
- The STIP core framework API gives the STIP-compliant applications access to the operating-system services and drivers in a consistent way (all devices are accessed using the same methods and calls). Also included is a standard API for activating applications (known as *stiplets* in the STIP world);

- STIP profile implementations: there are profiles (sets of services) for a generic card payment terminal, for a mobile telephone and for a FINREAD device (a secure PC-connected smart-card terminal designed for home use, e.g., for internet banking).

The device specification also includes a specification for the internal structure of a stiplet, dividing it into the *core logic component*, which is independent of the platform or environment in which it is run, and those services and functions that are platform- or application-dependent. Again, the idea is to minimise the amount of work needed when an application is transferred from one terminal type to another.

9.3.7   GlobalPlatform system specification

The final piece of the GlobalPlatform jigsaw is the system specification. This forms a toolkit that can be used to construct and manage a smart-card scheme; they include definitions of a smart-card management system (which I will address further in Chapter 12), the use of the ECMA-262 (ISO 16262) scripting language for loading applications, parameters and keys to the card, and a common personalisation system. The messaging specification is a part of the system specification that aims to define the rôles and responsibilities in such a way as to reduce interface development costs and facilitate collaborative software solutions. Also included is a set of functional requirements for a key management system, for managing both application keys and Global-Platform-specific card keys.

The comprehensive nature of these specifications means that few schemes will implement them in their entirety. However, each component does contribute to the openness of the scheme and the portability of its components, and where an application developer wants to ensure the maximum possible market for its application, or a card issuer wants to ensure that it can accept the widest possible number of applications, GlobalPlatform does offer a structure that is accepted as a standard throughout the smart-card industry and many key application sectors.

9.4   JavaCard-based products

There is a wide range of JavaCard-compliant products on the market, from all the main smart-card vendors. Some vendors have even implemented reduced instruction set (RISC) processors that execute Java instructions directly, thus reducing the size of the JavaCard API and improving the performance of the card.

The best-known implementations of GlobalPlatform are IBM's JCOP (Java Card open platform) card operating system, which has been licensed by many smart-card vendors, and the Prisma operating system, developed by Proton World International and now forming part of ST Microelectronics. Both implement the main features of JavaCard and the GlobalPlatform card specifications, but other products, such as KEB's Kona dual-interface card and NTT Corporation's ELWISE card, were the first to complete the new GlobalPlatform self-certification compliance programme.

Prisma includes a number of other GlobalPlatform-compliant features, including a component called CALC (card and application life-cycle), which follows the GlobalPlatform life-cycle management principles and which also includes standard interfaces to selected personalisation bureaux and a GP scripting interface. ST's Matrix Card Management product also follows GlobalPlatform structures, but uses a proprietary scripting solution.

In April 2006, JavaCard V2.2.2 introduced specific support for contactless and dual-interface products as well as additional cryptographic algorithms. This has particularly helped to spur the use of JavaCard in e-passports. The JavaCard protection profile has also been updated to match the new specification.

9.5   Status and implementations

In pure commercial terms, JavaCard and GlobalPlatform have been a considerable success. Because they represent the de facto standard for GSM SIM cards (other than low-end cards used for prepay only), over 1 billion JavaCards are in use,[1] accounting for 15–20% of all contact-based microprocessor smart cards. For GlobalPlatform, there are 100 million cards in issue, across over 30 schemes worldwide, and more than 1 billion GSM cards using GP technology for OTA (over-the-air) downloads.

As we will see, however, the convergence between smart cards and mainstream computing has been much slower to take hold; smart-card development remains largely in the hands of specialist developers. Although the JavaCard 2.1 specifications have been available since early 1999, and the Prisma implementation since 2000, the list of applications for which smart cards are used has changed very little since then. Most general-purpose Java development tools can be used for JavaCard applications, yet programmers responsible for security or portable data storage requirements on PC or server systems would be unlikely to choose a smart card unless it were explicitly specified.

Part of the reason for this is that the JavaCard programmer, or the GlobalPlatform scheme designer, must still be highly aware of the features and limitations of the card platform, terminal or card management system being used; the ability to use a general-purpose language does not mean that general-purpose features can be incorporated. The complexity still exists, even if more of it is contained within libraries and components that the programmer or implementor will buy in rather than resolving from first principles each time.

This suggests that there will remain a group of programmers and systems implementors who specialise in smart-card systems, although recruiting into that group should be easier than it was when every platform required a knowledge of its own machine code.

Implementing a full GlobalPlatform-compliant scheme using products from multiple vendors remains a complex activity. The emphasis on openness and

---

[1] At the end of 2005; source: JavaCard Forum.

self-certification demands a high degree of trust and mutual understanding between suppliers. In time, the range of GP-compliant products and applications will grow and this will make it easier to add an application to an existing scheme, or to change platforms within a scheme. With the very small number of cross-sector multi-application schemes that do exist today, some of the benefits of GlobalPlatform are not yet being realised. From the point of view of the multi-application scheme designer, however, the ability to use all or some of the GlobalPlatform concepts and specifications, and the extent to which those specifications offer options and scope for individual design and implementation decisions, offer probably the safest and most future-proof framework currently available.

9.6   References

[1] GlobalPlatform. www.globalplatform.org

[2] Sun Microsystems, Inc. *Java Card Technology*. http://java.sun.com/products/javacard/

[3] *Java Card Forum*. www.javacardforum.org

[4] Gosling, J., Joy, W. and Steele, G. S. *The Java Language Specification*. Addison-Wesley 1996

[5] Lindholm, T. and Yellin, F. *The Java Virtual Machine Specification*. 2nd edn, Addison-Wesley 1999

[6] Sun Microsystems, Inc. *Application Programming Interface for the Java Card Platform*. Version 2.2.2. 2005

[7] Sun Microsystems, Inc. *Runtime Environment Specification for the Java Card Platform*. Version 2.2.2. 2005

[8] Sun Microsystems, Inc. *Virtual Machine Specification for the Java Card^{TM} Platform*. Version 2.2.2. 2005

[9] GlobalPlatform. *Mission*. www.globalplatform.org/showpage.asp?code = missmstate

10 Multos

10.1 History

Multos was originally developed by Mondex International, the developers of one of the world's first electronic purses.

Mondex foresaw that issuers of electronic money would want to allow the e-cash application to reside on any smart card, not just ones issued by financial institutions. Hence, the electronic purse could find itself co-residing with other applications, such as a club membership or a credit–debit function, on the same card. It was also envisaged that applications on the smart card might be updated or added during the life of the card. And to protect the e-cash application from attack by fraudsters or hackers, it was essential that the multi-application smart-card platform must be secure and controlled by the card issuer.

Having worked on the electronic purse card for some years, the banks that owned Mondex could see the wide range of potential applications for the technology and, in particular, for the security functions it could offer. High levels of security, proof of security and tight control by the card issuer were key requirements. The design objectives of Multos were, therefore, fourfold:

- A very high security platform;
- A platform-independent programming language and application architecture;
- The ability for multiple applications to share card memory and data in a secure and controlled way;
- The ability to download applications after the card has been issued, but without the risk that unauthorised applications could be loaded or could corrupt existing applications.

Multos products first appeared in 1996, and in the same year the Maosco consortium was formed (by smart-card manufacturers and integrators, telcos and payment-card schemes MasterCard and American Express) to manage the specifications and to promote the operating system as an open standard. Mondex established a top-level Multos certification authority (known as the global key management authority or KMA), whose purpose was to provide issuers of Multos cards with a mechanism for controlling the content of the smart card, by providing a digital permission to third parties to install their application onto the issuer's cards. In January 1997,

MasterCard acquired a majority interest in Mondex International, and made it clear that the intellectual property in Multos was one of the strategic reasons behind the purchase.[1]

Since that date, Multos has gone through several releases to incorporate, in turn, updates to the EMV payment card specifications (for more detail see Chapter 15), PKI, contactless and ETSI (SIM card) interfaces, and new cryptography functions such as elliptic curves.

In November 2005, MasterCard, Hitachi, Keycorp and Oak Hill Venture Partners announced the formation of a new venture, called StepNexus, to further invest in and promote Multos. The aims of StepNexus are to exploit the patented, asymmetric, secure packet-content delivery mechanism that is used to install executable code and data to Multos cards by extending the types of trusted environment at which the mechanism can be targeted – such as new smart-card platforms or other trusted computing devices. In February 2006, StepNexus acquired HiveMinded, the developer of the Nectar SmartCard.NET platform for 32-bit devices (which is covered in the next chapter), and announced its intention to connect the Multos KMA to the .NET platform in addition to Multos.

The Multos specifications continue to be managed by the Maosco consortium, and StepNexus now operates the certification authority – in future to be known as the 'StepNexus Server'.

10.2   Scope and functions

Multos defines [1] not only an operating system, but also a language (Multos executable language: MEL), a card-management and application-downloading structure and a certification authority structure (operated by the Global KMA).

Figure 10.1 shows how this is structured; hardware, operating system and application developers are all licensed by Maosco, which ensures that the hardware and operating system are both certified to the appropriate security levels. Application developers deliver tested encrypted code to an application build-and-load process, which is controlled by the card issuer using certificates from the Multos KMA. However, the issuer never needs to see the application provider's code in the clear, nor share any secret keys with the application provider. This process, which is typically run by a bureau, delivers the application load units to the card. The card and its applications present an ISO 7816-compliant interface to the terminal. I will now go into some more detail on each of these functions.

10.2.1   Multos executable language

Multos executable language (MEL) is an interpreted language that was developed specifically for smart cards. It, therefore, includes only instructions and functions that

---

[1] MasterCard press release 1st January 1997, quoting Chief Executive Gene Lockhart.

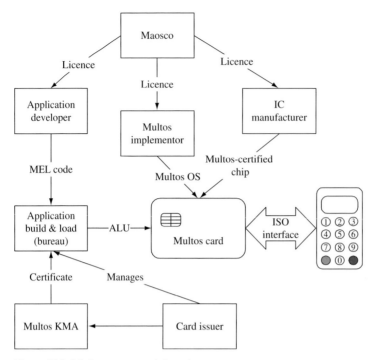

Figure 10.1 Multos scope and functions

are relevant to smart cards, although not all functions will be available on all cards. For example, not all platforms will support cryptographic operations such as check-sums or DES encryption as a machine code operation (primitive).

Although programs can be written directly in MEL, most Multos applications are developed either in 'C' or in the Multos assembler language. The Multos compiler converts them into MEL, which is the bytecode language used on the card. Programmers must be aware of the limitations of the smart-card platform they are using, for example its memory size, but they do not need to be aware of the structure of the card's hardware or operating-system functions.

10.2.2 Virtual machine

The MEL instructions are interpreted in strict sequence by a virtual machine known as the application abstract machine (AAM). The AAM is written in the machine code of the microcontroller; it sits on top of the hardware, and its purpose is to provide an environment in which applications can call services without knowing the details of the hardware or proprietary functions (see Figure 10.2).

The AAM provides each application with its own code space and data space (up to 64 kB each). The data space is further divided into static (non-volatile) memory, dynamic (volatile) and public (volatile) memory areas. The AAM also provides a number of mandatory and optional functions (primitives) that may be called up by the application.

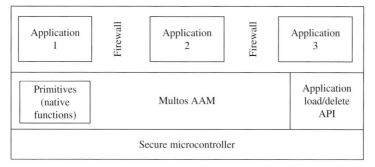

Figure 10.2 Multos application abstract machine (AAM)

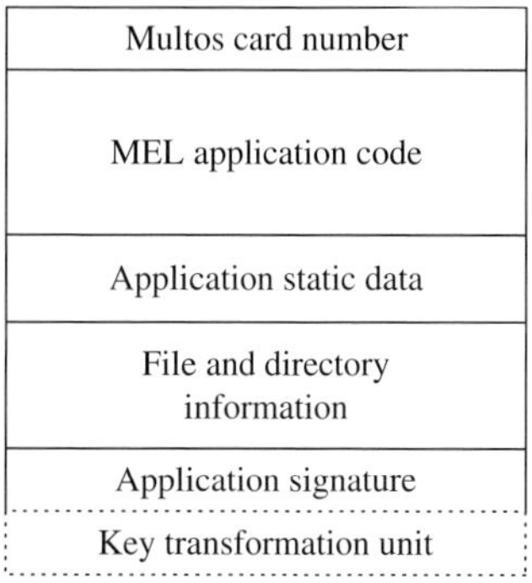

| Multos card number |
| MEL application code |
| Application static data |
| File and directory information |
| Application signature |
| Key transformation unit |

Figure 10.3 Structure of a Multos application load unit (ALU)

10.2.3 Application load and deletion

Following compilation, MEL bytecode is packaged into an application load unit (ALU) that also contains the card number where the application is to be loaded, any static data, the directory and file management information needed to store the application on the card and to allow it to be selected, and lastly the cryptographic data: a signature for the application to check its validity, and an optional key transformation unit (KTU) that allows the card to decrypt the application and static data (see Figure 10.3). If a KTU is used, then the whole application is encrypted into a single secure packet. This allows the ALU to be sent over any type of network, e.g., in an email, and stored offline for download to take place later. The advantage of this mechanism is that there is no need to establish a secure session with the Multos card. Secure sessions require reliable networks, otherwise they are vulnerable to sessions dropping and corrupting the card half-way through personalisation. The public-key encryption structure used by Multos enables the ALU to be sent in a single secure packet message, and protects the software without the need for security application modules (SAMs) or other intermediate devices: the code remains safe until it is decrypted within the card (and after that, by virtue of the secure loading and deletion structure).

In addition to the ALU (which can be created by the issuer itself, or a bureau or third party) every load operation requires an application load certificate (ALC), which is specific to the card and the application. Every application must be registered by the card issuer with the relevant KMA, which then generates an ALC for each card as requested by the card issuer. Most ALCs are issued by the Global KMA, but a small number of KMAs have been established by third parties, either for commercial or sovereignty reasons. For instance the Hong Kong and Saudi Governments and the Turkish military have established their own independent Multos key management authorities, so they have complete control over all keys that manage the content of cards that they issue.

The load procedure ensures that only the card for which an ALC has been generated can decrypt and load the application. There is a similar process for deleting applications.

10.2.4   Multos step/one cards

The normal Multos application load process described previously uses public-key cryptography; this technology imposes a floor for the card cost (the algorithm computations must be performed by a crypto-processor, otherwise the performance is too slow). Given the importance of the financial (payment-card) sector to Multos, it was important to have a card that meets the cost and performance requirements of that sector, which, in many cases, means a card that only supports symmetric cryptography (DES).

To meet this requirement, Maosco introduced the step/one card. This uses symmetric cryptography throughout, which keeps down the cost of both the card and the management systems, but, unlike standard Multos cards, is not intended to enable applications from multiple third parties on the same card without the need to share secrets. Application load certificates for step/one cards can be generated by the card issuer or bureau, which eliminates a further external interaction and possible cost. However, applications can still be downloaded post-issuance, and applications developed for Multos cards can also be used on step/one cards (provided they do not call any public-key encryption functions) and vice versa. In most cases the standard EMV application will be masked into ROM, which further reduces the amount of E^2PROM memory required on the card.

Although primarily developed for use in payment cards, step/one technology could also be used by several other sectors, such as public transportation or loyalty, which mainly rely on symmetric key cryptography.

10.3   Security

10.3.1   Common Criteria evaluation

Multos was designed from the outset as a high-security operating system. All Multos implementations are designed to meet the highest level of security evaluation, and many products are subjected to a formal evaluation following Common Criteria standards, or their predecessor ITSEC (IT Security Evaluation Criteria). As was mentioned in Chapter 3, the Common Criteria for Information Security (ISO 15408) are a framework for evaluating the security of a system or device in an objective and internationally agreed way.

A full Multos implementation should meet ITSEC E6 or its CC equivalent (procedures for evaluating to CC EAL7 are not available at the time of writing). These include evaluation of a formal model of the security mechanisms and a full inspection of the source code, as well as functional and penetration testing. However, the cost of preparing the documentation and submitting a product for these evaluations is substantial, and only products that include changes that impact the security enforcing functions must complete a full evaluation by a licensed laboratory.

The microcontroller on which the operating system is loaded must also be CC or ITSEC certified, in this case to EAL 4+ or its equivalent E3. Step/one implementations are not expected to undergo an EAL7 evaluation, but may be tested to the level appropriate for the environments in which they will be used. Typically this will be MasterCard 'CAST' or Visa Level 3 testing.

10.3.2 Program and memory management

On a Multos card, application code and data are stored in separate, predefined areas of memory. Multos guarantees that each application's data remain private and cannot be read or written by any other application. The exception to this is the public data space; an application may permit another application to access its public data after the first application has been completed or while it is suspended (see below).

Most application data and parameters in a smart card (other than purely transient or working data) are held in the E^2PROM. This has some special characteristics: it is written in pages of, usually, 32 bytes, and if the transaction is interrupted during a write operation then the page may be corrupted, including data that were unchanged during the write operation. Multos insulates the programmer from such worries and ensures the integrity of the data at the end of the transaction.

10.3.3 Runtime bytecode checking

The AAM verifies each instruction before executing it, to ensure that it does not attempt an illegal operation (in the same way as a conventional Java sandbox, but not as in a JavaCard). This gives an extra degree of reassurance that a new application can be loaded to a Multos card post-issuance, without needing to worry that it could be illegally accessing the memory space of an existing application. Applications may use their own data space or the public data space, and may also call up 'primitives' or machine-code functions (such as cryptographic operations), as this may be more efficient than having those same functions in MEL.

10.3.4 Shell mode

Most Multos applications will run in *standard mode*: that is, they conform to ISO 7816-4 numbering and application selection standards. However, a Multos card can also support a non-ISO application, by using *shell mode*. In this mode, all commands are routed to the shell application, so only one shell application may be loaded at any time. However, the shell application may delegate commands to another (standard) application.

10.3.5 Delegation

Multos provides a mechanism for one application to call another (for example, a common authentication or error-handling routine) without passing through the ISO application selection process.

The first application calls a primitive that suspends execution of the first and passes a command to the second. The second application may access public data belonging to the first and, when it has finished executing, it passes control back to the first.

10.4   Status and implementations

Multos is the benchmark for a secure, open card operating system. It forms the basis for the Common Criteria Smart-Card Protection Profile and all implementations have achieved certification at the highest security level (ITSEC E6 or CC equivalent).

Keycorp, DNP/Hitachi and Samsung provide a range of Multos and step/one products on Infineon, NXP, Renesas and Samsung silicon. As well as providing a platform for organisations developing their own applications, these companies offer off-the-shelf applications for EMV credit–debit, electronic purse and authentication. All the Multos suppliers have also developed dual-interface devices that allow contact and contactless applications to share a card, and that can be installed with an implementation of MasterCard's PayPass protocol application (which I will describe in more detail in Chapter 15).

Multos is available on cards with memory sizes from 8 kB upwards, and indeed is more attractive on larger cards (36 kB and upwards), where the overhead is insignificant and the memory and program management benefits are greater than on smaller cards. Multos has also been implemented by Keycorp on an NXP Pro-X chip, which offers MiFare™ emulation (see Chapter 6), for closer co-operation with transportation cards.

Despite its origins in the banking and payment cards sector, Multos is now having probably its greatest success in the government ID sector, with a number of major projects including the Hong Kong and Saudi Arabian national ID card and e-passport projects and the Turkish military ID cards. The strict hierarchical structure and provable security are strong selling points in these applications.

The introduction of Multos step/one technology has also greatly helped take up in the banking sector; in several countries there is a good business case for implementing multiple applications on a bank payment card, provided the cost of the card does not rise with every new application.

10.5   Reference

[1] *Multos Application Programmer's Reference Manual*. Maosco 1997–2005

11 Other operating systems

Every smart-card vendor has its implementation of the ISO 7816 file structure and application selection process: these form the vendor's native operating system and, as we have seen in Chapter 6, they can be used to create a multi-application card scheme.

To gain the full benefits of portability, a high-level language and post-issuance downloading of applications, you need a multi-application operating system like JavaCard or Multos. But there are several operating systems that sit between these two extremes, and which may offer advantages in some situations.

11.1 IBM MFC

IBM developed its first multi-function card (MFC) [1] in the early 1990s. It has since been developed to extend the cryptographic support and add new features. One of its most important features, however, is the ability to support applications in E^2PROM, which can be updated or downloaded after the initial issuance, using a scripting protocol.

IBM stopped supplying smart cards directly in 1999, but it has licensed the MFC to several manufacturers, and also develops tailored versions for specific schemes. It is now used by the French multi-application payment and e-purse card Monéo [2].

11.2 Advantis

Spanish card technology and systems supplier SERMEPA developed its 'TIBC' operating system in 1994 in order to run the 'Visa Cash' electronic purse product; TIBC was licensed to several card manufacturers and is still widely used in Spain and Latin America. Later versions of TIBC, now known as Advantis [3], were developed to support EMV (both Visa and MasterCard variants) and the CEPS electronic purse (see Chapter 15) alongside one another, in both contact and contactless form; this required public-key cryptography as well as the dual-interface capability. Advantis' independence from card manufacturers makes it a stepping-stone between the native operating system and a full interpreter-based system; however SERMEPA is a member of GlobalPlatform and future versions of Advantis will follow GlobalPlatform principles.

11.3   SECCOS

The Secure Chip Card Operating System (SECCOS) was defined by the German banking association Zentraler Kreditausschuss (ZKA), originally to support the Geldkarte electronic purse. It has since been extended to cover the strict security requirements [4] of the German electronic signature law [5], as well as the generation of one-time passwords (transaction numbers or TANs) for e-banking operations. The digital signature function is quite powerful and permits not only a legal signature for documents and emails, but also dynamic authentication for e-commerce and verification of certificates.

The key requirements of SECCOS are authenticated applications, public-key encryption support and evaluation against the Common Criteria Protection Profile for digital signatures at EAL 4 + or higher. Functionally, SECCOS makes use of the security environment and additional security commands listed in ISO 7816-8 and 7816-9.

There are three certified implementations of SECCOS: from Gemplus, G + D and Orga. Each of these offers a similar menu of applications: credit and debit card payment, Geldkarte electronic purse, electronic signature, TAN generation and a 'proof of age' function that can be invoked to prevent those under 16 or under 18 from purchasing products or viewing unsuitable content. Issuers can choose which of the applications they implement, but cannot add or delete applications. Parameters can, however, be downloaded or updated on the card using a scripting system.

A SECCOS card with dual interface (contact and contactless) has also been released; this also offers a MiFare^{TM} memory card emulation, to facilitate transition or convergence with existing transportation or access-card functions. The certified secure applications (debit card payment and electronic signature) cannot be used over the contactless interface.

SECCOS cards represent the state of the art for secure smart cards; they provide all the algorithms and authentication methods commonly used on cards, and support very long keys (up to 2048 bit RSA). There is, however, some penalty in performance and interoperability: a SECCOS card used in a high-performance terminal will give adequate performance for most transaction types; however when that card is used in lower-grade terminals (for example, in countries where only symmetric cryptography is normally used) transactions can take considerably longer, and may even fail if timeouts have been set too low.

This is a good example of the hazards of international interoperability: the standard of security demanded by the German banking industry and banking law is higher than in many other countries, yet German banks are required to protect their customers even when the transaction takes place in another jurisdiction with lower security demands but perhaps higher performance expectations. These hazards are likely to multiply when different sectors, each with their own security and performance standards, come together to operate a scheme.

11.4   .NET

Operating-system giant Microsoft announced its first foray into smart cards in 1999. Called Windows for Smart Cards (WfSC), its aim was to extend the Windows environment down into the chip card. This would allow the huge Microsoft developer base to start writing programs for chip cards using their existing languages and toolkits. However, the strategy required a specially designed chip because the operating system was overlaid directly on the hardware, and was not designed to use ISO 7816 protocols. WfSC was better adapted to the PC market than to existing smart-card markets, and as smart cards failed to achieve rapid penetration of mainstream PC applications, WfSC had not achieved significant market penetration before being dropped in 2001.

In 2004, Microsoft supported a new company entering the smart-card market with an updated version of this strategy. Microsoft has given the name .NET to its vision of a web-connected world, in which web services are available to all users on a network, connected by generic and universal standards, such as extended markup language (XML), simple object access protocol (SOAP) and universal description, discovery and integration (UDDI).

Two concepts that are central to .NET are those of *flexible client-server relationships* and *authentication*. Client-server structures are not as rigid or hierarchical under .NET as with more traditional models; a device may be a client at one moment and a server at another. However, a web-connected portable device such as a mobile phone or PDA is an obvious client. For it to use web services, a device of this type must authenticate itself to the server, and this is where smart cards become important. Smart cards may also be used to provide distributed security services or data portability in a .NET environment.

Many of the benefits of .NET will be familiar from the JavaCard discussion:

- Programmers can create applications using widely used languages and development tools;
- Multiple applications can be stored on the card and, indeed, may appear to run simultaneously (although in fact the card only responds to one command at a time, consecutive commands can be directed at different applications);
- The card application is unaware of the card hardware characteristics, and of the interfaces and protocols being used to transmit its data, be they ISO 7816 or more common PC standards such as TCP/IP or USB.

The interface between a .NET card and the other parts of a web-enabled network is through a *smart-card subsystem* that is a standard part of the Windows .NET environment. This subsystem enables devices and clients on the network that are not smart-card aware to communicate with the card, and vice versa.

There are, however, no fixed standards or specifications for the card itself or for the operating system. At the time of writing there is only one implementation of a .NET-enabled smart-card operating system [6], but it has been licensed to several smart-card vendors, making it the de facto standard.

In addition to implementing the relevant protocols, SmartCard.NET and its successors, Nectar and Hummingbird, provide some features such as caching for E^2PROM write commands, which considerably improve performance for those applications that spend time waiting for write operations to complete. Instead of the *common object file format* (COFF) used by other .NET applications, they implement a compact file structure more suitable for smart cards, and they also improve the efficiency of garbage collection (the process for clearing up after a .NET application has run).

The .NET card is being used to provide authentication in an X.509 environment (ANSI X.509 is the most widely used standard for personal digital certificates) and in some campus applications (notably by Microsoft's own employees [7]). But it has been slow to take off in other 'conventional' smart-card fields, nor has it yet been able to drive significant penetration of smart cards into mainstream desktop applications, such as e-commerce. The acquisition of HiveMinded, the company behind the SmartCard .NET family of products, by StepNexus, the owners of the Multos standard, may result in some convergence between these two product families.

11.5   Special developments

Smart-card technology and standards are now sufficiently widely understood that it is possible for a specialised software vendor to develop its own operating system, with all or some of the multi-application characteristics described in the last few chapters.

This tailoring of features may be appropriate where a specific combination of security and flexibility is sought: for example, a card issuer may want the ability to download software patches and parameters affecting its own applications, but not to permit any new applications to be downloaded to the card. Another issuer may want more flexibility for partners to download new applications, provided there is space on the card and it has tested and certified the application. A third issuer may see the ability to rent space on the card and to profit from transactions as a key part of its business model; in this case the ability to support standard languages and development tools is critical, as is the ability to record load operations and transactions and to transmit those to the card issuer host.

Examples of specially designed operating systems include:

- The Malaysian personal ID card, known as MyKad, which is described in Chapter 17;
- The card used by French banks during the transition from an earlier national standard (known as B0') to the international EMV standard – this card is derived from the IBM MFC.

Both of these cases involved large card volumes (over 15 million cards) and the investment would probably not be justified for smaller volumes.

Earlier in the history of smart cards, several companies (for example, in the satellite television business) deliberately sought to implement proprietary protocols and encryption algorithms to protect their intellectual property. In many cases, this turned out to be a poor investment, as the techniques were weaker than the well-known and standardised algorithms, and were broken within a few years by hackers. Most modern schemes have reverted to using publicly available algorithms and protocols, even where this has involved paying a licence fee. Nevertheless, as we will see in Chapter 14, satellite television is still an area where proprietary standards are common, even at the operating-system level.

11.6   Comparing operating systems with multi-application features

As we have seen, most smart-card operating systems provide some multi-application features, and for the organisation seeking to implement a multi-application scheme it is worth considering which of these features are essential, desirable, not important or possibly even undesirable.

These features include:

- **ISO 7816 file structures and application selection**: adherence to the international standards is likely to be essential if the cards are to be used in readers and terminals not directly controlled by the scheme owner or issuer.
- **The ability to store applications in dynamic memory rather than in ROM**: if the application is in ROM then it cannot be deleted or updated, nor can new applications be loaded. There is a specific technique used where the main application is stored in ROM but may be subject to updates or patches: this is to plant a large number of 'hooks' in the code that jump back to a control program stored in flash or E^2PROM. Any update, instead of the stored program, can then be called on by the control program.
- **The ability to download complete new applications or parameter sets**; the answer to this depends on a wide range of questions:
  - How long is the expected card life?
  - How likely is it that we will develop a new application during that time?
  - Could the new application be loaded now?
  - Who will control the new application?
  - How big might the application be? And, hence, how long might it take to load?
  - How easy is it to re-issue the card?

The cost differential between a simple (static) card and a dynamically downloadable card is often such that two or three static cards could be issued for the same cost as a dynamic card.

- **Support for high-level languages and development tools**: this is important if new applications will be developed frequently during the life of the card, or if the time-to-market requirements of any new development point to the need for rapid development.
- **Inter-process protection**: the need for this will depend on the extent to which applications can be developed and tested alongside one another.

- **On-card cryptography and issuer controls**: this will depend on the extent to which the issuer relies on the card to implement its security, or whether there are other systems in place that provide the necessary security.
- **Support for card management and wider scheme controls**: as we will see in the next chapter, a complex multi-issuer scheme that permits post-issuance application downloads will almost certainly need a smart-card management system with a wide range of functions and controls. Simpler multi-function schemes are much less likely to need these functions.

11.7   References

[1] Hamann, E.-M., Henn, H., Schäck, T. and Seliger, F. Securing e-business applications using smart cards. *IBM Systems Journal*, **40** (3), 635–47, 2001
[2] www.moneo.net/
[3] www.sermepa.es/ingles/index.htm
[4] CEN Workshop Agreement (CWA) 14168/9. *Security Requirements for Secure Signature Creation Devices EAL4/4+*. European Committee for Standardization 2001
[5] *Gesetz zur digitalen Signatur*, Article 3 of Informations- und Kommunikationsdienste-Gesetz (Multimedia Law) 1997
[6] *Smartcard.NET^{TM}: the Multi-application, Multi-language, Smart Card Platform, Executive Overview*. Corporate publication, Hive Minded, Inc. 2002–2003
[7] Hoffman, K. E. *Microsoft Employees Get Carded*. http://redmondmag.com/features/article.asp?editorialsid = 524

12 Card management systems

12.1   Legacy card management functions

Most card issuers will need some form of card management system (CMS), that allows them to keep track of the cards they have issued, expiry dates, etc. The CMS may also contain the details on the card, or may refer to another database (for example, a personnel database) that contains this information. The CMS may also include functions for maintaining the data (for example, name and address data), but for larger systems this is more often regarded as a separate customer management function.

For complex applications such as credit-card issuing, the CMS may link to several other systems, such as an authorisation system, call centre, statement and mailing management.

A CMS for magnetic stripe cards is usually a fairly simple 'flat' file structure, providing a link between the card number and the external data. With this kind of structure it is quite easy to give a call centre, for example, limited read-only access and the ability to make notes linked to the card or account, but they cannot affect transactions carried out by the card. The CMS can also act as the interface to a bureau or outside processor, so that the card issuer maintains the database but the bureau handles all the card-related functions.

12.2   Additional functions for smart-card management

When an issuer moves to a smart-card platform, it often assumes that it will need a smart-card management system (SCMS). This is not necessarily the case, initially at least – the term SCMS is used for a specific set of functions, and misinterpreting this can lead to high costs for managing an inappropriately-specified platform. It is, however, important to understand what type of CMS is needed, when an SCMS is appropriate, and how and when to migrate.

12.2.1   Basic smart-card issuing

Where the card application is static (will not change during the life of the card) and there are few card-holder-specific data items stored on the card, the basic legacy CMS may still be adequate, particularly where the card will be personalised by a bureau or external processor.

Even if personalisation (or the preparation of the personalisation data files) will be done in house, one can set up a template for the chip parameters and merge these with the card-specific information to give a personalisation file that can be sent to the personalisation bureau – see Figure 12.1. This is the solution that has been adopted by most bank-card issuers, since it minimises the initial cost and is easy to manage. Even for complex applications like GSM SIM cards or payment cards, there are off-the-shelf products available for personalisation preparation.

12.2.2   Links to real-time databases

Where the card management system links to a real-time transaction database (e.g., for authentication or authorisation purposes), this link is now more complex since the card may contain or use variable data: a transaction counter, or a record of stored offline transactions. We can either store these extra fields in the card management system or (often more efficient) store a pointer to the most recent transaction or other data record.

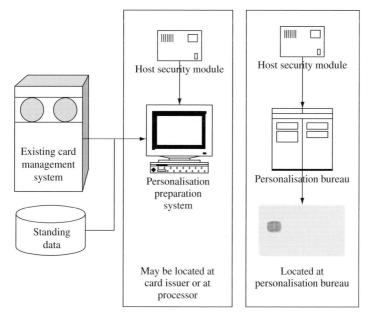

Figure 12.1 Personalisation preparation system

12.2.3   Scripts and parameter updates

Many smart-card schemes use scripts for updating parameters on the card. A script is a message sent to the card by the host, usually as part of a transaction response. For example, a transport card might, when passing through a gate, receive a message that its validity has been extended, or extra value has been loaded into the 'pay per ride' function. These messages must be very short, so that there is no delay to the original transaction, but they allow the card to update one or more fields in its memory, as instructed by the host. And the messaging system must normally be very secure, so that spoof terminals cannot intercept them or send 'increase value' messages to cards.

Again, there are off-the-shelf products available for script handling, which are designed to minimise the changes needed to both CMS and authorisation systems. The CMS may call up the script handling system as part of its transaction processing, or may contain a flag to indicate that a script is pending for that card.

The script handling system (see Figure 12.2) will store standard scripts and allow new scripts to be generated. Scripts can be scheduled either as a direct result of a transaction (e.g., a PIN change or value load) or by a batch system (e.g., a risk management or fraud detection system). And the script handling system will track the delivery of scripts, only cancelling a scheduled script when the next transaction has confirmed that delivery was successful.

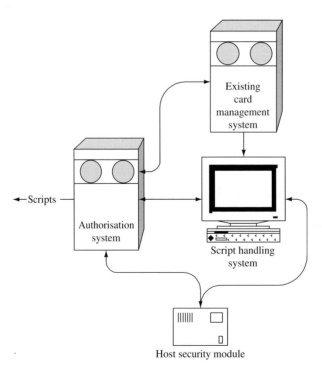

Figure 12.2 Script handling

12.2.4   Customer-relationship management and the 'segment of one'

One benefit of using smart cards (rather than magnetic stripe or read-only techno-logies, such as bar codes) is their ability to store data that not only vary from customer to customer but also evolve with time. This is potentially of great benefit to customer-relationship management (by taking account of previous transaction history in processing the current transaction) and to the development of a 'segment of one' marketing approach (since it allows products and messages to be tailored to the customer).

From a card-management point of view, however, the more data that are stored on the card, the greater the problem if that card is lost or must be upgraded or re-issued. There is a particular issue where any form of value is stored on the card: not only may customers expect that value to be protected, but in some circumstances the issuer may have a legal obligation to replace or honour that value, even if the card has been lost.

As the card issuer seeks to differentiate more finely within a given product, it becomes necessary to extend the number of fields in the CMS to hold these variable data. These fields may be accessed by other systems during transactions or for batch updates, as well as being used when the card is re-issued. This not only increases the size of the records in the CMS but also introduces a level of complexity into the management of access to these records.

A smart-card management system is designed to address this complexity, and will usually link to the card production system to ensure that any replacement card has the same features as the card it replaces.

12.2.5   Multiple application issuers

Having several static applications on a card does not usually increase the complexity significantly; the applications should all have the same expiry date (if needed). Each application will typically manage its own data; transactions are routed to different systems at an early stage by a switching system in the terminal or network. Sometimes a terminal will generate two messages for one underlying transaction, for example, one for the payment and another for loyalty points, mobile phone top-up or membership verification – see Figure 12.3.

The complexity does increase significantly, however, if the applications are owned or developed by different organisations, or if there are commercial or liability implica-tions that cross organisational boundaries. For example, if a card contains both a transport and a credit-card payment application, and the card-holder fails to settle the credit-card bill, then the bank will want to block the card, and may even retain the card if it is offered at an ATM. How is the customer to travel home? This requires the applications to be blocked separately. However, if the customer reports the card stolen, then both will want to block it as quickly as possible, regardless of which has handled the report of the loss.

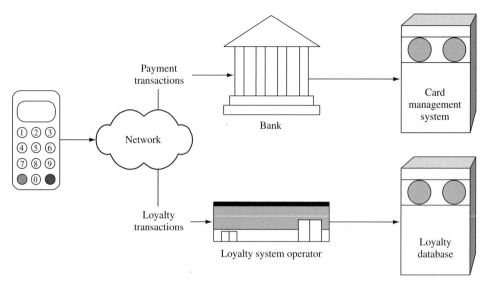

Figure 12.3 Routing multiple transaction types

Smart-card management systems are again designed to handle these different situations and may include interfaces to more than one call centre with different responsibilities and rights.

12.2.6 Post-issuance application downloads

It is not always possible to load all the applications at the time of initial personalisation: sometimes the core applications will be held in ROM but additional applications (dependent on the card-holder profile) may be loaded afterwards, before or after the card is issued to the card-holder. Any of these situations, where some of the applications may be considered 'live', should be treated as post-issuance download from a security point of view.

A card that offers post-issuance download is a much more flexible tool for the issuer, and may save considerable cost if the requirements change during the life of the card. But here, there is a requirement not only to ensure that any applications are restored to their correct state when a card is re-issued, but also to manage the download process itself, to authenticate any applications downloaded and to ensure that they are loaded into a suitable memory space on the card. As we have seen in earlier chapters, both GlobalPlatform and Multos place considerable emphasis on this aspect and both have smart-card management specifications that complement their card specifications.

An important part of this is the card's ability to recognise an application that has been approved for download; this requires a key management structure that allows the application issuer to show the card that the application is authentic and has been certified by the card issuer. If a symmetric key structure is used, then it is likely that the card management system will have to play a part in that structure (for a public-key structure, the card issuer's public key may be stored on the card for verification of the application's load certificates). And the card management system must also keep track of the available memory on the card, taking into account the need for temporary storage during the download and decryption process.

An SCMS is more or less essential for managing post-issuance download; if applications will be added to the card after it is issued, if those applications will be changed substantially during the life of the card, or if we want the ability to re-issue a card that contains multiple dynamic applications, then an SCMS will manage the complexity better than any set of subsystems.

12.2.7 Life-cycle management

A smart card is a computer system that will go through several states, from its first creation using an IC with a hard mask, through one or more stages of personalisation, issuance, blocking and unblocking, and finally expiry and destruction. As we saw in Figure 9.7, the applications on the card go through a shortened version of this cycle, not necessarily on the same timescale. (The one thing that should always be avoided is having an application on a card with an expiry date longer than that of the card itself, or of any master application.)

During the card production process, the card goes through a sequence of testing, writing chip and card manufacturer data to the card, and finally personalisation. At the end of each stage, a lock is irreversibly blown, and the card moves on to a new status. Depending on the relationship with the personalisation bureau and card manufacturer, it may be necessary for the issuer to store production-related data such as card and software version, batch numbers, freeze dates for changes, despatch dates or key identifiers for the personalisation files. These data can be made available to the call centre or to re-issuing systems, allowing more complex queries to be answered or cards to be re-issued at very short intervals. This requires an exchange of data with the personalisation system.

The smart-card management system tracks these stages of the life-cycle and advises the issuer when there are any conflicts or when action must be taken to restore a card to an operational state.

Whereas the legacy-card management system managed well with a flat-file structure, the SCMS has all the characteristics of a relational database. Figure 12.4 shows an example of an entity–relationship diagram for a typical SCMS structure.

12.3   Deploying a smart-card management system

While the basic technology of smart cards is well understood and deployed across a wide range of sectors, implementation of full smart-card management systems has been much more patchy, with some sectors making much more use of them than

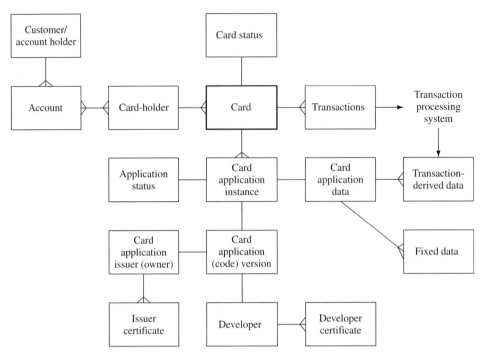

Figure 12.4 Entity–relationship diagram for a smart-card management system

others. Smart-card management system software has not been widely deployed by banks or public-transport issuers, and much of the experience in this area comes from sectors such as the corporate ID card, government and mobile telephone industries. Most SCMS software has been developed within one or other of these sectors, and subsequently generalised for the other sectors. So the first thing is to ensure that the system is fully 'aware' of the specific requirements and structures of the sectors where it will be used. It must interface with all the relevant customer records systems, real-time processing systems, delivery channels and front-end systems.

Across a portfolio of card products (particularly if some of them are issued to businesses and their employees) the relationships between cards, card-holders, accounts and account-holders may be quite complex. The complexity increases when we consider the applications, keys and parameters inside the card, other applications and, in particular, any external application owners. It is always worth drawing an entity–relationship diagram to map these structures, which will vary from issuer to issuer and from product to product.

It is also important to be able to define the uses to which the SCMS will be put – do we need, for example, to manage external application issuers or multiple certification authorities? Will application download be a regular or just an occasional requirement? What databases must we interface with? What parameters will be set on an individual card or card-holder basis and will these be adjusted automatically or manually? These answers may affect not only the configuration of the SCMS but even its selection.

One of the benefits of a multi-application card scheme is the flexibility to expand its functionality gradually; when the first cards are issued, the full final structure may not be known and hence it is difficult to answer these questions. On the other hand, it is important to implement an SCMS before the management of the scheme becomes too complex – in other words, it must be implemented before it is needed. To face this situation, many SCMSs have 'progressive' features or licence arrangements, which means they can be implemented in phases.

Lastly, the management of the SCMS must be planned and designed. This may be a substantial operational activity in its own right, requiring expertise across card and software technology, business requirements and partner or channel management.

12.4   Functions of a smart-card management system

A smart-card management system integrates all the functions described above:
- Storage of card-specific data fields, including fields that are dynamically updated by the real-time transaction processing system;
- Script handling for parameter changes;
- Downloading and updating card applications;
- Storage of data for multiple applications;
- Storage of production data and card re-issue data;
- Life-cycle management through creation, issuance, application or card blocking, expiry, etc.

Most systems will have some additional functions, for example:
- Key management for public keys and support of host-security modules (HSMs) for management of symmetric keys and certificate generation;
- Biometrics support (registration, template storage);
- Stock management for blank (unpersonalised) cards;
- Card-layout design and card-design storage;
- Small-volume badge production and emergency issuance;
- Audit trails and management reports.

This set of functions enables an SCMS to re-issue a card that looks almost identical to the original, without the need to integrate several new subsystems. There are many external interfaces, and so the initial integration effort is still significant; but changes and upgrades are easier to accomplish than with a legacy CMS or collection of subsystems.

Many sectors have specific requirements, as we will see in Chapters 14–18. The telecoms sector needs support for the GSM SIM and SIM toolkit specifications. Banks need support for the EMV specifications and risk management; they must also meet the payment-card industry (PCI) requirements for the security of stored data and transactions, and for access to card details. Many transport operators will need support for MiFare^{TM}, FeliCa^{TM} or other wired-logic cards. Government card projects may require a wide range of interfaces to different departments, each with its own access rights and security requirements. Some applications will require certification by a card scheme or security laboratory.

There is a specific group of products designed to meet the needs of smart-card-enabled secure physical and logical access, driven by the requirement of the US Homeland Security Presidential Directive 12 (HSPD-12), which requires all federal agencies to meet a national standard (FIPS 201 [1]) for personal identity verification. Although only mandatory for US government bodies, this standard may be regarded as a benchmark for secure access control.

Most suppliers have developed their products first to meet the needs of one specific sector, and then extended the functions to cover the needs of other sectors. For that reason it is important to ensure that the functions of any system being purchased cover the full range of likely requirements over the life of the scheme, or can easily be extended and certified.

GlobalPlatform publishes a comprehensive set of functional requirements [2] for an SCMS; although not a complete specification, it does describe the components and functions required to manage any GlobalPlatform card or application throughout its life-cycle. Many SCMS products (whether or not certified as compliant) support both GlobalPlatform and Multos application-loading and deletion protocols.

The following case studies give two examples of organisations deploying a smart-card management system, and show some of the lessons learnt in this process.

Case study A – King Fahd University of Petroleum and Minerals

by Erik Wellen, General Manager Professional Services, Bell ID and Dr M. Kousa, Associate Professor of Electrical Engineering and Director of Telecommunications Department, KFUPM

Organisation
King Fahd University of Petroleum and Minerals (KFUPM) is located in Dhahran and is regarded as Saudi Arabia's leading institute for science and technology, with over 10 000 students. The campus card solution secures access to the 900 acres of the University, 28 buildings and over 300 access control points.

The selected contractor was Bell ID, a leading player in smart-card management systems since the early 1990s, primarily in market segments such as finance, government and corporate ID. The subcontractors were Magna Carta (electronic purse – the Netherlands) and Bell Security (physical access control – United Kingdom).

Business requirements
King Fahd University for Petroleum and Minerals decided to implement a 'smart ID card' scheme on its campus in Dhahran, Eastern Province. The main requirements for the turnkey solution were the visual identification of the students and staff around the campus and physical access control, as well as closed e-purse. The system also had to be prepared for possible future extensions, such as an open e-purse, post-issuance capabilities and PKI.

Solution

Technology

The core system of the architecture is formed by the ANDiS Card and Application Management System, provided by Bell ID. This system takes care of the photo capture, biometrics enrolment and card personalisation. The University opted for two main card types: a contactless card for students and a hybrid card (contact and contactless chip) for faculty and staff. Both are personalised using a DataCard 150i card printer with integrated contact and contactless card reader/writer – three printers are installed on various locations around the campus. MiFare™ readers (supplied by BQT from Australia) are used for the biometrics enrolment as well as for building access. The access control system Pacom Graphical Monitoring System (GMS) provided by Bell Security takes care of room and parking access, controlling over 300 access control points on the campus.

The system architecture is portrayed in Figure A.1.

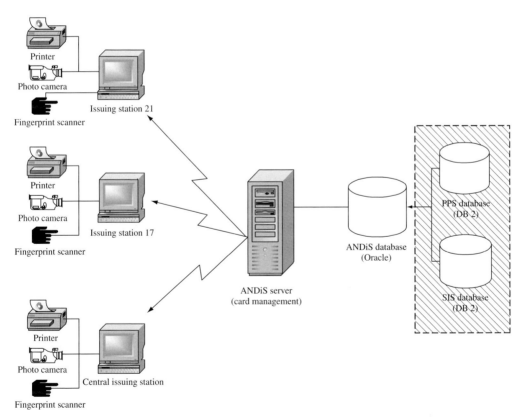

Figure A.1 System diagram for King Fahd University card management system (courtesy of Bell ID and King Fahd University)

Operations and organisation

When rolling out such a state-of-art campus card scheme, several new components and systems needed to be set up. The main systems were the smart-card management, access control and the electronic purse – all three had to be integrated and linked to the relevant existing systems in the University, namely the Student Information System (SIS) and the Payroll Personnel System (PPS) for faculty and staff.

On the campus, three issuing stations were set up to provide card enrolment and issuance services for (1) faculty and staff, (2) students and (3) others, such as dependants, alumni, visitors and VIPs. Each station is equipped with two terminals, web-based access to the ANDiS CMS-AMS (card and application management systems), biometrics equipment as well as a fully fledged photo studio.

Problems encountered

Although CMS is the core of the system, the project includes two other major modules: access control and e-purse.

For CMS, more detail and documentation of the client requirement would have made it easier and faster to optimise the design of the system (card types, user categories, card layouts, etc.). Linking several databases running on different platforms in a way transparent to the operators was one of the largest challenges.

The physical installation of the access control system was quite complex (civil work, cabling, networking, magnetic locks, etc.) especially for a site under full operation and normal daily functions. Intensive planning and co-ordination between various departments were essential for the success and timely completion of the project.

As for the e-purse, the main challenge was in motivating customers and cashiers to use it. Its success has very much to do with the culture of the community, and developing multiple services where they can be utilised.

Outcome

The smart ID card has been rolled out on campus and is well accepted by the campus users. Over 15 000 smart cards are in use on campus, providing user convenience to their card-holders.

Moreover, KFUPM has established itself as a centre of excellence in the field of smart cards in the Kingdom. Their system has been acting as a live, full-scale demonstration of the technology to other universities and organisations in the region.

Case study B – LG Card

by Jusuk Lee, Project Manager, IBM Korea

Organisation

LG Card is the largest credit-card issuer in Korea. Starting out in 1988 with 140 000 card-holders, it now has 9.7 million card-holders and 2385 employees. For its EMV card-management-system implementation it worked with IBM Korea; in addition to several EMV projects worldwide, IBM has implemented many employee ID and national ID projects and, as a full member of GlobalPlatform, was able to contribute specific card-management expertise.

Business requirements

LG Card wanted to:
- Centralise the management of its chip-card products, applications and stock of plastics. Although LG Card started to issue chip cards in 2004, by mid-2005 it had no organised process for managing its smart-card information, including white cards, card applets, card instances and keys. Without a centralised warehouse for this information, it was unable to gain an overall view of the card business.
- Make all personalisation modules re-usable by incorporating previous versions. Before the implementation of the smart-card management system, the personalisation modules and processes were not re-usable; for each new product a completely new module had to be created, which was costly and increased the time to market for new products.
- Establish a portal to support its card-holder servicing requirement. After an issued card is delivered to the card-holder, the bank may want to add, delete or upgrade individual applications on a card-holder-by-card-holder basis. To support all these diverse requirements, and to manage the history of each card issued, a post-issuance card-management system was seen as inevitable.

Solution

As a basic policy, LG Card requested that the smart-card management system (SCMS) should cover its legacy issuance environment, and should be compliant with GlobalPlatform. The IBM Smart-Card Personalization Manager was able to meet both of these requirements, managing both existing personalisation platforms and also the new issuing systems with a solution that had been developed exactly in line with GlobalPlatform guidelines, from card-component profile to system messaging.

LG Card purchased an additional host-security module to provide secure issuance facilities, and adapted its key management systems to generate, import, export and delete the keys used during personalisation and issuance.

There was a requirement to migrate about 800 000 chip cards, representing eight different card products, that were issued by the legacy system. The basic rule for data migration was that cards issued before the SCMS was installed should be available for exactly the same post-issuance services as the post-SCMS cards. For the migration, the IBM SCMS project team defined a data format, consisting of the mandatory data needed for post-issuance servicing. A pilot was carried out, covering data for tens of cards for each product and including a test of post-issuance downloads via the SCMS. The actual migration took several days to complete.

Problems encountered

There was a problem with the numbering system used for card image numbers (CINs). Within GlobalPlatform, the CIN is the issuer's unique reference, used to identify an individual physical card; this is linked to the card reference number (CRN), which is the logical reference but may be applied to several physical cards. Before the introduction of the SCMS, LG Card had been issuing cards without recording which products were being used. During the project, LG Card and the IBM project team had to establish a CIN generation rule that not only covered cards issued under the SCMS, but also created a numbering scheme for cards issued under the legacy systems. To solve this problem, the CIN was constructed from product information and a unique sequence number.

Outcome

The SCMS project started in February 2005 and was completed that September. In addition to the 800 000 IC cards that were migrated from the legacy system, a further 1 million cards were being managed by the SCMS in June 2006. After an interim phase of parallel operation using both the legacy and new SCMS systems, LG Card now manages all of its chip cards through the new SCMS.

12.5   References

[1] *FIPS 201: Personal Identity Verification (PIV) of Federal Employees and Contractors.* National Institute of Standards and Technology February 2005
[2] *Multi Application Smart Card Management Systems: Functional Requirements.* GlobalPlatform May 2001

Part III

Business requirements

13 Common business requirements

Having looked at the technology available to create a multi-application card scheme, we now turn to the business requirements – what are organisations trying to achieve when they launch a multi-application scheme? What are their underlying aims and how do these differ from sector to sector? If several parties are co-operating on a scheme and they each have different aims, this may cause problems from the start.

In this chapter, I will consider the requirements that are common to all multi-application schemes, and in the next five chapters I will look at some key sectors that have implemented smart-card schemes. In each case, there is scope for combining applications within one sector or across sectors: combinations within a sector may be difficult for reasons of competition or service scope, whereas across sectors the issues may be cultural or organisational. I will explore these in more detail in the last few chapters of the book.

13.1 Card issuing

Issuers generally have one of two motives for issuing a card: the first is to provide a simple *record of entitlement* (proof that some money has been paid or that a person may enter an office). This type of card is often short lived and it is probably a simple data carrier, adding little value. Typically, the requirement for such a card is to provide the record of entitlement at the lowest possible cost consistent with an adequate level of security and reliability. Such cards are very unlikely to carry multiple applications.

Other cards are issued with the aim of creating a link between the issuer (the bank, telco, bus company, employer or local government office) and the end user. Here there is much scope for adding value and deepening the relationship by adding applications or functions to the card.

The value of such a *relationship card* to each of the parties depends greatly on the reasons for the relationship – commercial, practical or legal – and whether it is competitive or unique. This affects many of the characteristics of the card-issuing process and the criteria for an optimum process and so, in turn, points to potential problems, where there are different types of relationship between the end user and the various application issuers or service providers within a single scheme.

13.1.1   Differentiation

Card issuers in competitive markets are likely to want to differentiate their products as far as possible; they may do this through the features and functions offered on the card or through the card design, or the differentiation may have nothing to do with the card, but only allow access to a differentiated service. The range of parameters and data storage available on a smart card allows service providers to tailor each card, and the services it offers, to each customer: the high-spending customer may be offered a wider range of features and services than his or her twelve-year-old child, or the second card may include a number of features for the child's protection. The existence of multiple applications on the card is, itself, a form of differentiation: each customer need only have or see the applications that add value to him or her.

In some other situations, however, such as provision of government services, visible differentiation – which could be seen as discrimination – should be avoided at all costs. Strict standards of uniformity must be observed. The card-management-system requirements for such a scheme are much simpler than for a highly differentiated scheme.

There is not necessarily a conflict between these two requirements, but it is important to understand where each component fits on this scale and how the requirement for differentiation is squared with the need for uniformity and standardisation.

13.1.2   Issuer control

Most card issuers feel strongly the need to control all aspects of the cards they issue, from the graphics and quality of the plastic, through the mailers and communications with users, to the software stored on the card and any parameters that the software uses. This can still apply to a multi-application card, although issuers should understand that, as with PC software, it is not possible to guarantee that any smart-card application will be 100% free of errors or unexpected behaviour.

When, however, several service providers decide to share a card, possibly each using its own card software supplier, one or more of them must accept significantly less control than if they each issued their own card. This is why organisations such as GlobalPlatform and Multos attribute specific rôles to the *card issuer* and *application issuer*. Usually, the card issuer has full responsibility for the selection of the plastic, the card operating system and the core application, as well as working with the card vendor and any personalisation bureau (I will come shortly to the operational aspects).

One very important, and often emotive, aspect is control of any *branding* on the card. Every brand owner is very protective of the way its brand is used and how it sits with other brands – a glance at any payment card will show a distinct hierarchy of bank and card scheme brands and acceptance marks, each in a precise size and location on the card. Any other brand (such as a co-issuer or an acceptance partner) that must also feature on the card upsets this carefully negotiated balance. And what if the colours of each brand simply clash? These considerations can be sufficiently contentious to make many forms of card-sharing impossible.

13.1.3 Interoperability

I have used the term 'card scheme' quite freely through this book without ever defining it precisely. It has both commercial and technical implications: at its core may be a set of rules, a card and terminal specification or both. A simple card scheme such as a company's ID badge may not need any rules, but many schemes involve a number of *card issuers*, as well as *acceptors* who accept the cards and provide services.

Unless the cards and terminals are all of the same type and owned by the same organisation, a card scheme must find a way to show that all cards – regardless of their issuer – can communicate with all terminals, regardless of the acceptor: this is the key to interoperability at a technical level. There are still business rules to determine whether the card will actually be accepted: a Paris métro ticket may be read by a London bus but not necessarily accepted.

A successful card scheme will ensure (through accurate specification and thorough testing) that all cards can communicate with all terminals, and that the business rules agreed by the participants will be followed, with no unexpected behaviour or conditions arising.

13.2 Card and card-holder management

Any card scheme must have some form of card and card-holder management system. As discussed in Chapter 12, the requirements will vary, ranging from a relatively simple cross-reference file to a formal smart-card management system that encompasses all the applications on the card.

Where there are several independent applications on the card, decisions must be made as to the extent to which card management is centralised: if transactions are handled by separate systems, each may need access to some card-management information. If several service providers are involved, then it is likely that each will need some form of card-management database to manage the applications under its control.

13.3 Application development

Several factors drive the optimum choice of application development tools and channels:

13.3.1 Speed and cost

For some applications, the speed to market is very important. In this case, it is a big advantage to be using a high-level language with good development support tools and emulators available. This will also usually reduce the cost of developing the application but it may increase the amount of memory involved, compared with a native code application. At the other end of the scale, some applications or functions are so common that they can be written in native code and programmed into the hard mask, or ROM memory, of the card – this will usually result in the lowest-cost card but may make subsequent changes very difficult or impossible.

13.3.2 Future-proofing

During the life of the card, not only may new applications be needed, but also bugs may be discovered or specification changes may require a small update to the program. To protect against these, some development tools and operating systems can allow programs to be segmented and partial updates or patches to be downloaded.

The cost balance must be calculated: is the overhead that allows applications to be updated in the field greater than the cost of issuing a new card? Would it indeed be possible to update all cards in the field? If all cards do need updating, could this be done within a reasonable timescale through re-issuing the cards or would a field upgrade be more practical? These questions can only be answered with knowledge of the operating environment as well as the respective card and development costs – a mobile phone SIM card is much easier to update and maintain than an infrequently used local transport concession pass.

13.3.3 Platform independence

Many issuers wish to avoid being dependent on one single manufacturer or card type: they may want the freedom to negotiate prices with several suppliers, as well as ensuring security of supply in the event of a plant failure or supply chain breakdown. Independent application developers also prefer to develop code that can be used on several products from different manufacturers.

In practice, no smart-card application offers complete platform independence, but some allow a wider range of choice than others. Many manufacturers do have some form of cross-licensing agreement with their competitors, while other operating systems are available for several card platforms. In particular, JavaCard-based products offer a much wider range of compatibility across different manufacturers and memory sizes than any native code or operating system. This allows issuers to source products from several suppliers and to change their suppliers from time to time.

13.4 Application and memory management

Applications loaded on to the card are subject to a hierarchy of control: the application developer is responsible to the application owner. Where that owner is not the card issuer, the issuer must normally retain ultimate control, but may not need access to the detail of the application, only its external characteristics and interfaces. Applications are granted the right to use memory by the card issuer, or by the terminal owner in the case of terminal applications.

Where there are both contact and contactless interfaces on the card, the card issuer must also control whether applications are accessible from the contact or contactless interface, or both. Unless the card uses two chips (one for the contact and one for the contactless interface), in principle any application could use either or both. However, there may be practical, psychological and security issues, for example:

- Contactless readers are generally used at public-transport gates, but most of the readers installed in retail terminals offer contact only;
- Contactless transactions are generally much faster than contact, not only because of the higher transmission speeds but also because there is less need to place the card accurately in the reader: however, there is a higher rate of failed transactions (which must be repeated) and this costs time;
- Transactions that take more than a few hundred milliseconds are likely to be too slow or unreliable if carried out over a contactless interface, since the card may move out of range;
- Card-holders may be worried about 'phantom' transactions or information being read from the card without their knowledge – this risk is probably exaggerated but does exist;
- Transactions that may include scripts or parameter changes may require the card to be retained in the reader for longer than normal, and are therefore often better carried out using the contact interface.

13.5   Terminal management

Many of the arguments and criteria discussed for cards also apply to terminals and terminal applications. As we saw in Chapter 7, modern terminal applications are likely to be developed in high-level languages and can relatively easily be transported from one terminal type to another. Most terminals are placed for one primary purpose, and that core application changes very little, if at all, during the life of the terminal. However, other applications may be added or changed much more frequently, and in some cases – particularly supplementary terminals in retail or kiosks – the mix of terminal applications is quite volatile.

The need for dedicated terminal management systems or subsystems is now much more widely accepted, whether the terminals are counter-top units in retail outlets, public-transport gates or mobile phones. In some schemes the responsibility for terminal management is quite dispersed, but it is good practice to have an overall structure and hierarchy of control similar to that described for card applications. As described in Chapter 7, the functions required may include management of transaction collection and hardware maintenance as well as applications.

13.6   Operations

Every organisation that wants to issue cards must make a fundamental decision as to how much of the card-issuing process will take place in house and how much will be outsourced.

For magnetic stripe cards, even quite small-scale issuers could, in fact, do everything in house, using a PC and a desktop card printer-encoder. However, both security and economy of scale do play a part, and when card issuing reaches a scale that requires a department of its own, many issuers decide to outsource all or part of the activity to a bureau or third party. The parts of the operation that are most commonly outsourced are *fulfilment* (sending cards to end users), *personalisation* (encoding, printing and sometimes embossing), and *card management*. A card-issuing bureau will often have physical and cryptographic security processes and approvals, ISO 9001 procedures, back-up facilities in case of failure, and possibly a means of issuing emergency cards at short notice.

For a multi-application card the primary issuer will normally dictate the criteria. What level of security is required? How fast must customers receive their cards, on initial application or if the card is lost? What is an acceptable level of cost for card issuance (it is often several times the cost of the card itself)?

13.6.1   Customer perspective

Customers and end users of multi-application cards have a common need for *clarity* of the relationship between them, the issuer and any service providers, their responsibilities in respect of the card, and to whom they must turn for assistance in the event of any problems. Much of this is good practice for any card scheme, but for a multi-application card, where there is scope for considerable confusion in customers' minds, it is essential. There is a need for a single document that explains, 'This is how you use this card,' rather than a series of documents relating to each application or service. It is particularly helpful for issuers to print the relevant telephone number or website address on the card, to avoid the card-holder guessing which service provider to turn to in case of problems.

When using the card, customers also need to understand what *functions* are being used – not from a technical point of view, but as a user and a data subject. For example, what data are being passed to the retailer or service provider? Some cards ask the user to select an application (for example: credit, debit or electronic purse); in these cases it is important that the meaning of the choice is clear to the user (and not only to the bank).

There are accepted *standards of service* in each sector and country for handling problems with a card or replacing lost or damaged cards. However, these standards vary widely from sector to sector and even from country to country. For example, many GSM operators can cancel a SIM card and issue a replacement immediately, over the counter, where others require a new account to be opened. In the UK and USA a lost bank card is normally replaced within 24 hours, whereas in some countries a delay of 7 days or more is quite normal. These differences may matter greatly if the card is the only way a person can get to work or pay for medical treatment.

13.6.2 Retailer/acceptor perspective

Retailers and others who process card transactions have a similar need for clarity as to relationships and responsibilities; they are also often concerned about the liability they may incur, particularly for services and events outside their control. They, too, need to understand what level of service they can expect in the event of terminal or transaction problems: many issues are caused by helpdesk operators who are unable to understand the problem because they do not understand the acceptor's business, or even the language they are speaking.

Many card transactions have to drive or connect to other functions within the acceptor's own systems; in this case a clear interface specification is critical, and in most cases also some assistance in implementing the interface (through provision of software, hardware or consultancy).

13.6.3 Back-end systems

The business requirements for back-end systems will largely replicate the standard requirements for systems in that sector: an appropriate balance of standard (preferably open) platforms and functions, support availability, reliability, flexibility, ease of modification, cost, etc. As I discussed earlier, the actual functions to be performed in house will also vary according to the needs and priorities of the business.

For a multi-application card, very often the requirements will be similar or identical for all the applications on the card. It is probably not serious if the card issuer has slightly higher demands in one area than other application owners. However, if it is the other way round (the card issuer's systems do not meet the requirements of another application owner) or if explicit trade-offs have to be made that do not suit all parties, then we can again suffer from a mismatch of requirements – this can happen between departments within one organisation as well as across companies or sectors.

13.6.4 Exception handling

Almost every card operation expends 90% of its effort on the 10% of customers, cards, terminals and transactions that are in some way exceptional: customers changing their details, lost and stolen cards, terminals that fail or become unreliable, or transactions that are the subject of a commercial dispute. The efficiency of a card operation is largely determined by the speed with which these exceptional cases can be resolved and the degree of automation that can be applied to them.

In most established card operations, a long history of experience in handling these transactions has resulted in efficient workflow processes and decision-making rules. Moving to a multi-application card requires these processes and rules to be reviewed and new ones to be drawn up for situations that involve more than one application.

13.6.5 Cost and revenue management

Whether the card operations are a cost or a profit centre, some form of cost management should always be in place; the revenue side is often omitted, but for many card schemes revenue management may be just as important. For a multi-application card, it may be quite difficult to attribute costs across applications, but a formula should be devised, even if the internal customers are departments within one organisation rather than separate businesses, to allow costs to be managed by all the relevant businesses and departments.

Many current schemes have insufficient management information (MI) to manage the business successfully. Although larger schemes in sectors with a strong tradition of cost control are likely to be better than smaller schemes where the card is merely a small part of a bundled cost, there are often surprising gaps in even the best-run schemes. It is important that the management information requirements be defined at the outset so that the necessary fields can be included in messages and data structures.

13.7 Security

A final requirement common to all card schemes is security: while the security requirements for a bus discount or student attendance card are quite different from a military security pass or a bank's master key generation card, each represents a security policy that must be defined and implemented using the card.

A smart card offers a very flexible way to implement a security policy: each user can be given specific rights, and these can be enforced using physical and logical access controls. Transaction authentication can be performed using either symmetric or public-key cryptography, and all transactions can be logged and traced to a specific card or user.

A multi-application card can form part of several security domains, helping to enforce the security policy of each. Common security functions such as user authentication (by PIN or biometric data), message authentication and transaction logging can be implemented on the card and shared by several applications.

It is worth noting that the security offered by any card scheme is not only dependent on the card and the transactions it supports, but is also affected by the strength of the registration and card-issuing processes, and by the physical security of the manufacturer and issuing bureau. This is an area where different sectors will often have quite different standards, but no application can offer security stronger than the weakest link in this chain.

13.8 Trust and liability issues

Any card scheme depends on a degree of trust between the players, even if the security policy is there to reinforce that trust and to provide the tools to enforce it.

A multi-application scheme demands a higher level of trust among the main players (in particular, those who will share the card) and between them and the acceptors who rely on the information in the card. Good fences make good neighbours, however, and the scheme will always work best if each party's responsibilities are clearly set out and understood.

A specific query with a multi-application card scheme is often: what happens if party A relies on information about party B that has been provided (stored on the card) by party C and this information turns out to be wrong or out of date? These situations can only be handled by the contracts and rules of the scheme, which will usually limit the liability of the information provider, however it is always best if such situations and disputes can be avoided altogether by the design of the scheme and the data flow between the parties.

13.9   Special needs

One application or function that may apply to any issuer in any sector is the ability to recognise and make allowance for people with special needs. Almost half the population has a special need of some kind, and probably half of those needs have some effect on the way these people use cards and card systems. The most common set of needs is visual impairment: a long-sighted person who needs glasses to read but might not wear them when out shopping, or a person who simply needs more contrast on an outdoor display in broad daylight. Shaky hands, the need for one hand to hold on when standing, poor memory or slow reading speed are also common problems.

Sometimes, particularly in response to laws or bad publicity, service providers make specific adaptations for one type of special need. However, they must be aware that specific adaptations can make life more difficult for another group: for every very short person there is a very tall person who finds low terminals unusable. System designers' preconceptions of the problems faced by, for example, elderly users (who often need many minor adjustments) are often very different from the perceptions of the users themselves. It is much better to ensure that for every process there is an alternative or alternatives suitable for those who, for some reason that we may not have envisaged, are unable to use the process.

Well-designed card systems can make life easier for many groups of people; however this will only be true if the ergonomic design has been carefully thought through, not only in the context of the target user group envisaged but for all those who may use the system.

There is a set of European standards (EN1332 [1]) that covers several ways to make cards and card-based systems easier to use for those with special needs. It includes a notch to help insert cards the right way round, a recommended layout for keypads and raised symbols on keys. EN1332-4 is a standard for encoding special needs onto chip cards; it defines the holder's needs in areas like character size, screen brightness, audio prompts and extended timeouts. This could be added to almost any multi-application card (whether it uses a native or multi-application operating system) and would both increase usage of the cards and help to dispel resistance to cards from many special-interest groups.

The draft standard EN1332-5 is of particular interest for multi-application cards as it standardises tactile identifiers for different applications, allowing a visually impaired user to distinguish between a transport card, a bank card, a medical card and one that has several functions.

13.10   Reference

[1] *EN1332: Identification Card Systems – Man–Machine Interface.* European Committee for Standardisation 1999–2006

14 Telecommunications

14.1 Telephone cards

The very first smart cards were issued in the mid 1980s as disposable prepaid cards for public telephones. They replaced magnetic stripe, optical or inductive cards with a technology that was generally more reliable; over 95% of all telephone cards for automatic use now employ smart card technology. The competition is not from other card-based systems but from centralised systems where the account is held on a host system rather than on the card itself (even if a card is used to deliver the account number to the user).

The business requirement for a public telephone operator is to eliminate the collection of cash in telephones using a reliable system with a minimum of moving parts. Coin-operated telephones are expensive to build, since they must be very robust; they are expensive to operate (the cash must be collected regularly) and to maintain (coin mechanisms frequently become jammed or are vandalised).

Smart cards can be sold by retailers like any other goods and so can be made very widely available. The 'float' of value sold but not yet used is available to the operator and can earn interest (although it does still represent a liability in accounting terms). Smart cards, therefore, offer a very convenient and portable way to sell value; however they also have some disadvantages:

- The cards themselves must be manufactured and distributed; card cost may be as low as 8–10 cents[1] but as a proportion of the smallest denomination value (which may be $2 or even less) this is still substantial. In addition, the distributors and retailers need to make a margin, which may be 5–15% of the card value;
- The cards represent cash value, even when they are sitting on a retailer's shelf; there is, therefore, a cost for stocking the cards, and a significant risk of cards being lost or stolen in the distribution chain;
- Most telephone cards are disposable, to avoid the need to build a network of loading terminals and the risk of unauthorised reloading; however not only does this mean that the card has a relatively short life, it also gives the operator very little information about its customer.

Many operators are, therefore, moving towards the 'calling-card' model, in which the account is held on a host system and may be topped up by the user, using Interactive

---

[1] 'Cents' are used to represent either dollar or euro cents where the values are approximate.

Voice Response (IVR) or a similar system. In this case, the card itself need carry no value and, indeed, a paper 'scratch card' can be used. Although this is arguably less convenient for the customer, it can be quite versatile and the value can be used in different ways; in particular it allows 'virtual networks' where the network operator does not own the telephone or other infrastructure.

The first generation of telephone cards used memory cards with an issuer authentication area, logic to prevent the value on the card from being increased, and a relatively small maximum count (100–200 bits). Subsequent generations have increased the maximum count (up to several thousand bits, allowing other services as well as better granularity), greatly improved the card authentication mechanisms that protect against counterfeiting and emulation, and added a user memory area, allowing storage of numbers called.

Most operators now use one of two authentication schemes, known as Eurochip and T2G. Earlier versions of both schemes have been hacked, but in their current form (Eurochip III and T2G +) they provide an adequate level of security for this application. In principle, operators using these schemes can share cards: telephones and other services can use a Security Application Module (SAM) to store the keys for each issuer whose cards they wish to accept, and this allows them to authenticate these cards. In practice, this facility is little used: it is easier for an operator to require customers to buy its own card than to accept other operators' cards under a revenue-sharing agreement, and customers seem to accept this.

14.1.1   Other payment cards used in public telephones

If a public telephone can accept a smart card, it could, in principle, accept other payment cards as well as prepaid cards. In practice, an operator wanting to accept bank-card payment transactions faces several obstacles: the telephones must be certified by the bank and card scheme, rules must be developed for managing liability for fraudulent transactions, and a protocol must handle the cases where the card is removed before the call or payment transaction is complete.

These obstacles were overcome by the French national payment card CB, working with France Télécom, and by the Belgian electronic purse scheme Proton. But few other payment card schemes, and no international cards, have crossed this bridge successfully. It would certainly be easier to use a multi-application card carrying both a bank payment application and a telco prepaid or account-based application.

14.2   Mobile telephony

14.2.1   Subscriber identity modules (SIMs)

Cellular telephony, which was introduced to most European countries in the early 1980s, expanded very rapidly in the late 1990s and early 2000s; at the end of 2006 the number of mobile subscribers was over 2.2 billion, of which almost 1.8 billion[2] used a GSM (Global System for Mobile Telephony) service and nearly all the rest used one of the CDMA (code division multiple access) standards, as used in the United States, Japan and Korea. The GSM standards were originally developed by an industry group, but are now owned and maintained by the European Telecommunications Standards Institute (ETSI).

All GSM telephones use a subscriber identity module (SIM), which is a smart card in the ID-000 format described in Chapter 3. The SIM is located inside the telephone and its rôle is to act as the agent of the network operator. Whether the telephone is bought from the operator or directly from a retailer, and whether the user has a contract with the operator or operates on a prepaid basis, the SIM authenticates the telephone to the network. The key business requirement for the operator (who owns the SIM and controls it at all times when the telephone is switched on) is to provide this authentication and to secure the network against intruders and eavesdropping (earlier analogue networks – the 'first generation' – were prone to both).

The operator's network authentication centre (AuC) uses its secret key and its own version of a set of algorithms, known as A3, A8 or A3/8, to set up a number of 'triplets' including a random number, an encrypted version of that number, and a key K_c that will be used for encrypting communications. The SIM card contains the same secret key and algorithm, and so can generate the cryptogram and K_c when fed the same random number. These triplets are distributed to 'location registers' belonging either to the 'home' network or a 'visiting' network to which mobiles may roam – see Figure 14.1.

Each time a mobile attempts to set up a call by communicating with a base station, the base station sends it a random number from this pre-stored list. The SIM encrypts this number and the phone returns the encrypted value to the base station, which compares it with the stored cryptogram and, if the two match, allows communication using a further, shift-register-based algorithm known as A5 and the key K_c. The A5 algorithm is now considered a somewhat weak link but the security is adequate for most voice and broadcast services.

However, the SIM also provides the operator with its only real opportunities for differentiation and additional services: the SIM contains the parameters that control the subscriber's use of the network, including any 'roaming' rights (the ability to use a different network), so that the 'away' network does not have to contact the 'home' network for these details. And it may be used by the subscriber to store telephone numbers and other data.

The first SIMs were invariably associated with 'post-paid' accounts (the subscriber paid on receipt of a monthly bill) and this limited the market to those with an established credit rating. The mobile telephony market took off when the prepaid business model was introduced; however the account was still held on the host system and the functions of the SIM were unchanged in this business model. Some operators (for example in China) have now moved to a variation on this model, where the SIM itself is associated with an account but represents a fixed value and is disposable when

---

[2] Source: Global Mobile Suppliers Association.

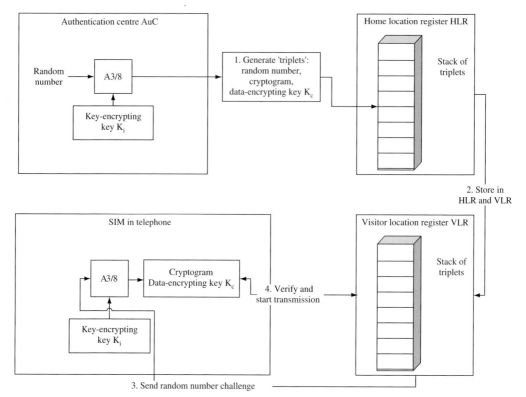

Figure 14.1 GSM authentication

it runs out. This avoids the need to run a top-up network but increases the cost and risk of SIM distribution.

The SIM forms a link between the operator and its customer; as such it offers significant opportunities to build on the customer relationship. However, these opportunities are constrained in many countries by data protection and competition regulations, so that it is very difficult for an operator to build a generic model that can be used in many countries. Operators in some Asian countries, where data protection regulation is weaker, have been able to store more data on the SIM and to use it in more innovative (and sometimes intrusive) ways, by sending unsolicited messages and even downloading new features to the phone.

14.2.2 SIM toolkit

In 1999, ETSI introduced a further facility called the SIM toolkit (STK). This provides a standard interface to the telephone's display, keypad and communications functions from applications running in the SIM. Data and applications can also be downloaded to the SIM; these are then selected by the user or activated by a command sent over the air, thus adding features to the telephone. The SIM toolkit has been used for presenting a menu of information services (such as traffic news), for location-based services (finding a petrol station or coffee shop in the vicinity), and for ordering goods; it is a versatile tool that considerably extends the scope of a standard 2G telephony service.

14.2.3 3G

The third generation (3G) of mobile telephones offers more data-focused and multimedia-focused services, together with higher data-rates and more flexible security mechanisms [1]. The competing technology camps have converged partially (though still not completely) so that handsets will be able to roam between all 3G operators, and although the use of a smart-card USIM (universal subscriber identity module) is not mandated, most if not all handsets will, in fact, offer one.

The USIM provides much of the flexibility in a 3G telephone: it enables roaming between 2G and 3G networks, provides the ability for a telephone to act like any other device on an IP-enabled (internet protocol) network, and plays a major part in the security access and authentication mechanisms that underpin the multimedia functions of 3G, as well as enabling functions such as virus protection, user profiles and downloading control. Its rôle as a service differentiator is, thus, even stronger than that of the SIM card.

It also updates the security scheme used by GSM: the 'triplets' issued by the AuC become quintets, with a network signature and sequence number, and stronger algorithms are used throughout.

14.2.4 Application and parameter updating

Both SIMs and USIMs are becoming larger; whereas in 1996 most SIMs were 8 kB or smaller, memory sizes for USIMs start at 64 or 128 kB, and continue to grow. A high proportion of modern SIMs are built on JavaCard platforms (although we must not forget that there is still a substantial market – particularly in China – for very small, simple SIM cards that are often disposable).

The larger SIMs are not only able to store additional applications (using SIM toolkit and similar techniques) but also allow completely new applications to be downloaded over the air, at any time when the phone is switched on. This gives the mobile operator an unprecedented level of control over the SIM, and hence over the behaviour of the telephone and the service offered to the customer.

The most advanced operators are now looking seriously at SIMs with capacities from 4 MB to 1 GB, and capable of supporting communications speeds up to 2 Mbps; we saw in Chapter 6 that these sizes and speeds are feasible with new memory types and construction methods. These devices can be used for storing substantial video sequences as well as card applications. A SIM-based web browser is likely to be a standard feature of these devices; the SIM will cache pages and predict loading sequences in order to improve performance.

14.2.5   Near Field Communication

In Chapter 3, I described the Near Field Communication (ISO 18092) standard for short-range contactless devices, which only works at very close range. This offers significant advantages for many transaction-based applications: transactions can only take place with the knowledge of the user and there is much less need for contention management or protocol negotiation than with, for example, Bluetooth. NFC has attracted considerable interest in the mobile telephone world because it offers the possibility of using the handset as a general-purpose terminal or intermediary device in many applications. Antennae and chips for NFC are usually fitted on the battery cover or case of the mobile phone – see Figure 14.2 – but it is technically possible to integrate them into the phone itself.

In 2005, Japanese mobile phone operator NTT DoCoMo and the East Japan Railway Company (JR East) started a large-scale trial that allowed their mutual customers to use their telephone handsets to reserve and buy tickets, to pass through turnstiles and to pay for goods in outlets that already accept JR East's Suica card. The handsets are fitted with an antenna and chip using FeliCa^{TM} technology, and the chip can also be accessed (using a contact interface) by the telephone. So for contactless payment the phone works like a contactless card, but for those transactions requiring the use of a keypad, display or communication with a host system the second interface comes into play.

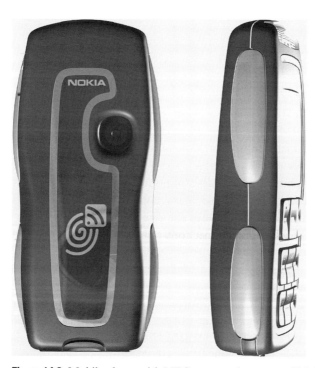

Figure 14.2 Mobile phone with NFC antenna (courtesy of Nokia)

Other applications envisaged for NFC include:
- Quick access to functions such as favourite URLs, message sending or even making calls;
- Open (contactless) payment, not only for transportation but for any goods, using a credit or debit-card function on the phone's smart-card chip;
- Card-holder authentication using PIN entry or other functions on the telephone;
- 'Active advertisements' that allow NFC-enabled devices to download further details of products, or even to buy them.

In all cases, the attraction for the mobile phone operator is the rôle played by the telephone in each of these transactions; it makes the phone a near-indispensible tool rather than an optional accessory and allows the operator to add value to the transactions. It may also open up the possibility of charging the transport operator, bank or advertising company a licence fee or transaction charge.

In some of these applications the mobile phone is a 'pivot' technology that forms the link between different technologies and networks, communicating in NFC mode with a card reader (and hence with another card) and in GSM/3G mode with a host system. A phone in a small retail outlet can be an acceptance terminal for customers and a payment instrument for the shopkeeper.

But operators must also be aware of the risk that such developments, although considered 'cool' and appealing by several user sectors, may not necessarily bring extra revenue and may in some cases distract from the business of driving transactions and traffic that will actually add value to the network.

14.2.6 Average revenue per user – the drive to add value

The economics of operating a telephone network are driven by the huge scale of investment required to set up the infrastructure and to recruit customers, the long timescales involved and the intense competition between a small number of players in each highly regulated market.

Internet-protocol telephony, internal networks and direct price competition are undermining voice telephony as a source of revenue. Customers are less and less likely to stay with a single network for long periods, and the cost of setting up a new customer (including, in many cases, a subsidised handset) can easily eat up the first six months' revenue. Customers who use prepayment rather than a contract have lower setup costs, but the cost of collecting revenue is higher. In many Western countries, there are more telephone connections than people, and so a network can only gain new customers by wooing them from other networks.

In both fixed and mobile telephony, voice calls are becoming a commodity: each operator can offer equivalent quality and the only way they can compete is on price. The average revenue per user (ARPU) falls continuously for any given service. It is, therefore, imperative for operators to find additional, chargeable services where the value to the customer is high in relation to the cost of provision, and which have some 'stickiness' or entry barriers so that customers cannot move to another network as easily as for voice calls.

Adding value at a network host (for example, by storing payment card and transaction details) is one way to do this; handset features such as NFC may offer another. However, each of these only works if the operator has some unique features that cannot be copied by competing networks. An exclusive partnership with a local monopoly or dominant provider is, therefore, often a characteristic of such deals.

14.2.7 Network–manufacturer relationships

Those unfamiliar with the mobile phone market must also be aware of an important strategic driver in this market: the battle between handset manufacturers and network operators for 'ownership' of the customer – or more accurately for primacy in the customer's mind.

Handsets are, in most cases, more differentiated than networks: people are likely to select handsets that fit their self-images and budgets, while most networks offer more or less equivalent facilities and functions. Handset manufacturers can offer software and other services that help to bind the customer to their products.

Networks do also have several cards to play. Through the SIM or USIM they can provide functions on the phone tailored to the user's needs, and because they are in constant touch with the user, they can be more aware of changes in the use of the phone or other customer requirements.

It would be destructive to the industry for either side to win this battle completely, and a balance is emerging in which network operators are specifying more features directly to the phone manufacturers. Independent software houses are creating software for network operators that can be loaded onto any manufacturer's phone.

Service providers and card issuers from other industries who wish to work with the mobile phone sector must remain aware of this balance and not, for example, work exclusively with a single phone manufacturer to develop a product that must be sold to network operators.

14.3   Mobile payment

One of the challenges for an operator seeking to increase ARPU is how to be paid for any extra services, without incurring high risks. If the extra service is a ring tone, weather forecast or other high-margin, low-cost service, then the best route is likely to be by charging to the customer's bill or deduction from a prepaid value.

As the value of the goods ordered rises, and delivery is less directly in the control of the operator, for example, a book or bunch of flowers, it may be more appropriate to offer an account that can be repaid or guaranteed using a bank direct debit or credit or debit card. For an air fare or higher-risk item, an actual credit-card transaction is probably the only satisfactory route. Network operators are very skilled at managing risk in their own environments; their back-end systems include complex and effective revenue management systems. But for the wider field of payments only banks and specialised payment service providers have the history and skills to manage their exposure satisfactorily.

So there is a need for services that bridge the gap between mobile telephony and bank payment services. There have been several initiatives aimed at solving this problem, most notably the Simpay consortium set up by the major European operators Orange, Telefónica Móviles, T-Mobile and Vodafone. However Simpay, like most of its predecessors, foundered after a key player withdrew. A more successful business model is that of NTT DoCoMo's iMode service, which does not attempt to link with external payment services but channels all payments through the phone's billing channel.

In the early 2000s, several operators supported the development of dual-slot mobile telephones, with one slot for the SIM and the other for a bank card, employer's card or other token. As customers moved towards smaller handsets, this became a less attractive proposition, and the emphasis has moved back towards making more intelligent use of the SIM card itself. NFC has now put the last nail in the coffin of the two-slot phone; it now seems much more likely that a phone would use contactless communications with any external card.

One of the very few services that actually places debit-card or credit-card functionality on a SIM card is SK Telecom's Moneta service (see Case study C).

Case study C – SK Telecom's Moneta service

Organisation
SK Telecom (SKT) is the largest mobile telephony operator in Korea, with over 16 million users and a market share of 53%. Its mobile payment efforts started in 1996; in 2002 it established the m-Finance Division and set up a co-operative structure with Visa International to develop standards for mobile payments.

Business requirements
SKT was facing slowing growth in its voice services and wanted to address this using a new business model that merges telecommunications and financial services [1,2].

As well as providing mobile payment, SKT believes that Moneta can be a new customer relationship management tool, enabling tailored communications and services to customers based on standing (customer) data, preferences (drawn from the transaction history) and location-based information.

Solution
The Moneta system comprises three components:
- A Moneta phone;
- A special SIM containing personal and financial information; and
- A 'dongle' or infrared receiver that receives messages using the Visa IrFM (infrared financial messaging) standard.

A customer making a purchase in a shop presses the telephone's Moneta key, enters a PIN and points the phone at the dongle; the credit-card data are transferred via

IrFM to the retailer's payment terminal. For offline (e.g., internet) purchases, the retailer sends an SMS message to the mobile; the card-holder enters his or her PIN and the data are passed by GSM to the retailer's host.

Initially the service used a separate card with the financial and personal information but this limited the number of phones that could be used and so was less attractive to customers. The Moneta SIM solution is more attractive but does mean that each SIM is associated with one bank.

Problems encountered

SK Telecom had to overcome many obstacles to launch this seemingly simple service:

- Convincing regulatory bodies: in addition to approval from the telecommunications authorities, SKT had to obtain security approval from the Ministry of Finance. The Ministry was concerned about the security of the wireless communication, the rôle of a telecoms company in providing a financial service, etc., and even after the security was demonstrated there was still a substantial delay in gaining approval. Ultimately, SKT requested new laws to handle this emerging service, and the government tabled a new Mobile-Financial Regulations Act to address this issue.
- Co-operation with financial institutions: at the time Moneta was launched, many banks were still backing two-chip solutions. SK Telecom also had to negotiate a share of the acquiring commission on transactions; as a result of these two factors the partners at the launch were two smaller card issuers rather than the largest players. To make the service more attractive to card issuers, SKT has now separated the ownership of the SIM from its maintenance, using security 'domains' on the card; card issuers can, therefore, own the SIMs while SKT maintains them.
- Technology standards: while SKT adopted the IrFM 0.56 standard, rival mobile operator KTF (a division of Korea Telecom) lobbied for government adoption of IrFM 1.0 as a standard. SK Telecom successfully argued that this was not an area for government standardisation.
- Penetration of dongles and Moneta phones: in order for this service to be attractive to merchants and users, there must be a critical mass of both. SK Telecom has decided that all new phones will include the necessary technology, and it is pushing for major retailers to install the dongles.
- Customer acceptance: many customers expressed security concerns, but these have reduced as the service has become more widespread, and in particular as the two competing mobile phone operators have launched their own services (incompatible with Moneta).

Outcome

After a slow start, SKT is seeing some expansion of Moneta usage, with over 1 million users and 500 000 terminals. It has extended the services covered by Moneta to cover travel passes, mobile stock exchange and mobile banking.

References

[1] Lee, H. G. *et al.* MONETA services of SK Telecom: lessons from business convergence, experiences for ubiquitous computing services. *Proceedings of the Second IEEE Workshop on Software Technologies for Future Embedded and Ubiquitous Systems* 2004

[2] Shin, B. and Lee, H. G. Ubiquitous computing-driven business models: a case of SK Telecom's financial service. *Electronic Markets*, 15 (1), 4–12, 2005

14.4   Satellite and cable television

Digital satellite and cable television using the digital video broadcast (DVB) standards [2] includes the ability for operators to encrypt channels using a conditional access (CA) system. There are several proprietary implementations of CA, all of which use a chip card to store keys that can be updated or cancelled over the air; the keys are used to decrypt the channels in real time for viewing. The CA card must remain in the slot all the time to allow updates to reach the card.

However, the DVB standard also specifies that set-top boxes should carry a second smart-card slot; although technically 'undefined', this was always envisaged as carrying a payment card or similar card that could be removed for security purposes – see Figure 14.3. This facility has been little-used, although for motives similar to those of the mobile operators, television network operators have become increasingly keen to deliver profitable added-value services in addition to entertainment and information. In some cases this has involved working with a bank to provide a payment service linked to a debit or credit-card account, as described in the case study below.

Figure 14.3 DVB receiver with two smart-card slots (courtesy of Pace Micro, Ltd)

Case study D – SkyCard

by Dave Taylor, Barclaycard Technology Office

Organisation

In 2002, British Sky Broadcasting sought a partner to issue a Sky-branded credit card. In addition to normal 'chip-and-PIN' credit-card functions, Sky wanted the partner to provide a mould-breaking, industry-leading, interactive credit card, with a market leading loyalty offering.

Barclaycard was selected as the provider and the two companies set up a partnership to issue and operate the card on a profit- and cost-sharing basis.

Business requirements

Sky's objectives were to increase the attractiveness of its service and its average revenue per user (ARPU), by providing over-the-air (OTA) functions including loyalty-point gathering and redemption, the ability for customers to view their card account histories and manage their accounts, and the flexibility to add services, channels and content providers to the card.

Barclaycard was keen to promote its existing EMV-based chip-and-PIN credit-card product and services as far as possible; it wanted a product that would extend its capabilities but that did not require large investment in new IT security structures and there should not be a requirement for data synchronisation or key management structures spanning the two organisations.

Solution

Constraints

All BSkyB set-top boxes have a second chip-card slot (the first is for the conditional-access subscription card): this had not previously been used to support payment cards (indeed, it had rarely been used for any purposes). Since this slot did not meet the EMV low-level electrical and protocol specifications, changes were required both to the card's chip and to the set-top box firmware to allow the card to be used in this slot.

Several enhancements also had to be made to the browser in the set-top box to support the second slot and the security model adopted for the service. There were also memory and performance constraints, with over 100 different variants of set-top box in the field.

This project was not happening in isolation, and during its development many changes were made to the electronic programme guide (EPG) in the set-top box, the WAPTV browser and the host systems in both organisations.

Technology

The design (see Figure D.1) required three applications on the card:

- A standard EMV application for credit-card use;
- A reduced EMV application to provide card-holder authentication for use with the set-top box and OTA applications;
- A 'data-object application' that provided a scratchpad memory that could be used by Sky and its content providers, and by the customer, with a wide range of access permissions appropriate to each class of data.

The card selected uses a native operating system and 4 kB of dynamic memory. Because the data-object application performs very little processing, it was found to be quicker and more flexible to use the native card functions than to use Java or another high-level language.

Architecturally and operationally, the key-management system was designed to ensure an unambiguous partitioning of security responsibilities for both organisations. Keys and their functions are controlled using an agreed set of key life-cycle policies.

Software (plug-ins) was developed to allow communication to the second card slot. The plug-ins were designed to expose an abstract layer to the chip card through the WAPTV browser, allowing content providers to develop chip-based applications more easily. Certain security functions, such as calls to the PIN functions, are restricted and controlled. The design, development and testing of the plug-ins were a considerable and complex part of the project.

All data from the STB to the Barclaycard system are encrypted end to end. This required a specific cryptographic component to be developed on the Barclaycard host system using PKI to deliver the required security – where signed public keys are sent through the broadcast stream.

Implementation

Following agreement of the contract, the service was launched in April 2005.

Testing was a lengthy process, not only because of the large number of set-top box variants, but also because the service had to be compliant with the open-access requirements of Sky's regulator, Ofcom, and be certified by Sky Subscribers Service, Ltd (SSSL). This involved a significant amount of resources to test and re-test the application plus ensuring that test slots booked were able to be met by the development teams.

Card production was kept as simple as possible by modelling the data required to be written onto the chip to be managed by the Barclaycard host system, rather than having to synchronise data from both organisations.

The chip had to go through additional rigorous security certification above and beyond EMV. Barclaycard was able to draw on the assistance and expertise of MasterCard's security employees and processes.

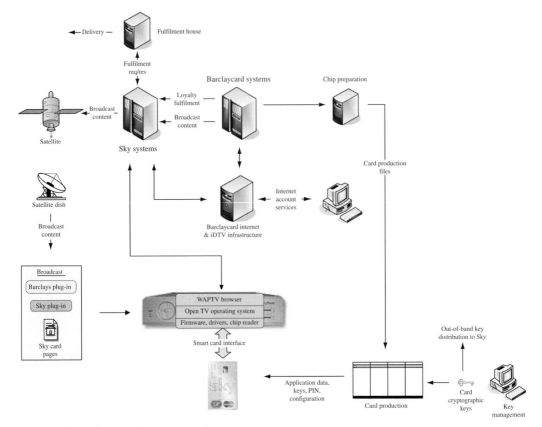

Figure D.1 SkyCard system diagram

Outcome

At June 2006, there are nearly 300 000 SkyCard customers.

The project has been seen as successful not only because it has exceeded its targets, with very few technical or customer service problems, but also as a demonstration for both companies of the ways that card or set-top functionality can be extended. Developing the business and IT requirements in parallel and having a single integrated cross-organisational programme ensured that the system, once designed, could be delivered, while the structure of the partnership meant that the objectives of both businesses were aligned. The SkyCard has won several industry awards for technology and innovation.

14.5   Internet services

Although we will see in several succeeding chapters examples of smart cards being used for logical access control, payment and user authentication, there are few examples of cards being used on a large scale for facilitating internet access.

This is perhaps surprising: even moderate business internet users quickly accumulate dozens of favourite sites, accounts and passwords, and spend considerable amounts of time re-keying frequently used data. Smart cards provide a secure way to store URLs, passwords and commonly used data, most of which are otherwise written down or stored in an insecure location. Access to the data on the card may be through a biometric or single password, which the user is forced to change regularly.

There are smart-card products available commercially to address this, and several banks now offer software or card-based solutions in order to protect their internet banking customers. But internet service providers do not seem to offer this service, and only relatively high-end business security packages offer smart-card-based single sign-on. This is a relatively straightforward piece of technology to deploy, and the reason for its scarceness is probably that no organisation wants to take on the possible support costs, or the liability if someone deploys it badly.

14.6   The future of multi-application cards in telecommunications

The telecommunications sector has been the engine driving the growth of the smart-card industry and has been responsible for much of the technological innovation in the field, particularly when it comes to downloading applications and managing cards in the field – key functions for multi-application use.

The card's ability to help identify and authenticate remote users puts it at the core of many services. Increasingly, however, the additional functions performed by the smart card, whether in a mobile phone, satellite receiver or internet service, are unrelated to the telecommunications service itself, but are driven by content providers or other partners who make use of the network to deliver services to customers. This suggests that telecoms companies may increasingly play the rôle of infrastructure providers rather than themselves managing additional dimensions to their customer relationships or benefiting directly from the added-value services offered by the card. This may fall to the other sectors, which I will discuss in the following few chapters.

14.7   References

[1] Hillebrand, W. (ed.) *GSM and UMTS: the Creation of Global Mobile Communications.* Wiley 2002

[2] www.dvb.org

15 Banking

Cards are now the payment instrument that banks prefer customers to use for spontaneous transactions. They are gradually replacing (in some countries, have replaced) cheques and other paper instruments for these transaction types, and even have a rôle to play in many regular or business transactions.

Using a card *removes a manual process* (capturing the transaction details) and with it the scope for errors that any manual process offers. Cards help customers to use *lower-cost channels* (such as ATMs and the internet) and, in the case of smart cards, also actively help in managing many *risks*: by forcing the user to authenticate him- or herself, by managing transaction 'velocities' – i.e., the rate of spending – and by offering the merchant or acceptor the opportunity to verify the identity of the bank or its membership of a valid card scheme. In some cases the card acts as the *agent* of the bank, authorising or cryptographically signing transactions on the bank's behalf.

There is also a *marketing* and *psychological* side to the bank–card-holder relationship: the card carries the bank's brand and is the permanent reminder in the customer's pocket. In a competitive business, where customers may have accounts with many banks, the card that is 'front of wallet', i.e., used most often, identifies the bank that has raised the level of its relationship from pure account-holding to that of a preferring customer.

15.1 Types of card

15.1.1 Credit, debit and charge cards

The best-known payment cards are those that either perform transactions directly on a current account (debit cards) or a special credit-card account that can either be linked to a current account or operated independently, with the card-holder choosing when payments are made. Included in the latter are charge cards, used primarily for travel and entertainment purposes.

Credit cards allow the financial institution (either a bank or an authorised lender) to earn money from the interest charged, but there is also a risk of non-payment. Most debit and credit cards are issued under the brand of a payment scheme (e.g., Visa or MasterCard for international cards, or a national brand) and follow a 'four-party'

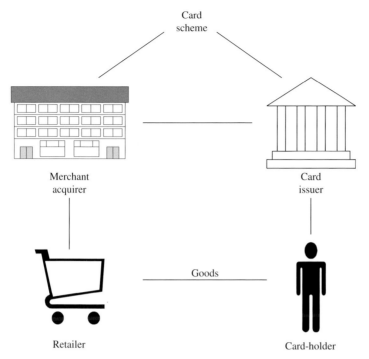

Figure 15.1 Four-party model for card payments

structure (see Figure 15.1). Under this structure, a system of *interchange* allows the issuer to charge a fee (usually a percentage for credit cards or a fee per transaction for debit cards) to the acquirer; the acquirer in turn charges the merchant. This way, all parties have a business incentive to increase not only the number of cards but also the volume of transactions carried out using the cards.

Regulators in some countries have objected to these structures, and in particular to the way interchange is determined (for international cards this is at a regional level unless there is a local agreement within a country). However, few regulatory interventions to date have helped to increase competition and investment (and some have decreased it), and so the basis for interchange is likely to remain. In some countries, banking law limits the extent to which banks can charge customers for cards or transactions, in other countries competition.has this effect; this means that there is always a degree of cross subsidy between services in this area and all charges must be viewed as part of the package of services offered. As with telephony, this provides an incentive for banks to increase the total value of the service package, and extra services on the card or terminal are often seen as a way to achieve this.

15.1.2 ATM cards

Most debit cards can be used at ATMs, often for operating the account (e.g., making transfers) as well as drawing cash. Credit cards offer less functionality at ATMs:

usually only cash withdrawal and PIN change. Within each country, one or more ATM networks link the ATMs of different banks; the facilities available at other banks' ATMs may be less than at the card-holder's bank's ATMs (known as 'on-us') but will generally include cash withdrawal and balance viewing.

'Not-on-us' ATM transactions are generally subject to a *negative interchange* régime: the ATM owner charges the issuer a fee for each transaction. Again, competition or banking law often restricts the extent to which this fee can be passed on to customers. However, many countries are now seeing a growing number of fee-charging ATMs, where the owner charges the customer directly for the transaction rather than the card issuer. In both cases, it is beneficial for the ATM owner to make its ATMs as attractive as possible to customers, to maximise revenue. Modern ATMs use internet protocols and graphics interfaces as well as offering advertising and as wide a range of services as possible.

To maximise both the range of services that can be offered to customers at ATMs and the number of customers who can use those services, many banks have, in the past, also offered ATM cards that do not have more general payment facilities. These more limited-function cards can be offered to a wide range of customers and help to reduce the bank's risk. With chip-based debit cards, this distinction is no longer necessary, since the functions and risk can be managed on the card, which varies its function according to the type of terminal.

15.1.3 Electronic purses and pre-authorised debit

While credit cards are 'pay later' and debit cards are 'pay now', the introduction of smart cards opened up the possibility of prepaid cards or *electronic purses*, also called *stored-value cards* in North America (in Europe this term is only used for closed systems such as a vending-machine card or travel card). With an electronic purse the value is stored on the card itself; transaction values are usually limited and the purchase process is kept very simple (no PIN or signature is required in most cases).

Banks are attracted to electronic purses because they offer the possibility of extending the bank's scope into much lower-value transactions than with debit or credit cards: transactions such as a newspaper or coffee purchase, or even an online article costing a few cents, could, in principle, be made using this medium. At a national level, cash handling accounts for a significant proportion of banks' costs, and cards offer a way to reduce this cost. And having access to information on these transactions helps banks to understand better their customers' needs and behaviour.

The 1990s saw the launch of many electronic purse schemes, many at a national level (such as the Danish Danmønt or Portuguese PMB) but also schemes like VisaCash and Mondex that were offered worldwide. In practice these schemes suffered from two main drawbacks:

- Transaction times were usually felt to be slower than handing over cash, because a contact card was used and simply inserting and starting up the card takes several seconds;
- A separate transaction was required for loading the card, reducing the convenience and incentive for the customer to use the card.

In many cases, the economics did not work either: the security requirements demanded either a very secure and expensive card or an online transaction, while a special terminal was usually required. It is extremely difficult for a bank or processor to find a way to charge adequately for processing low-value transactions.

In the mid 2000s two changes have taken place: *contactless cards* offer much faster transactions, while the introduction of *pre-authorised debit* transactions not only reduces the need for separate load transactions but also allows most terminals that handle debit cards to process e-purse-like transactions. A pre-authorised debit card does not store value directly, however it does keep a record of a small amount that has been pre-authorised by the bank and can be deducted from the current account without further authorisation. This process manages risk; it keeps the transaction flow very simple and, for the most part, transparent to the customer and retailer (except when a further authorisation is required).

In many situations where they would previously have considered an electronic purse, banks are now offering cards using pre-authorised debit, usually with a con-tactless interface. There is still one drawback: when the pre-authorised amount runs out, a full debit transaction (usually using the contact interface and with PIN card-holder verification) is required.

15.1.4 Prepaid and gift cards

Another variation on this theme reflects an updating of the prepaid card idea: instead of supporting a current account, the card can be run from a pre-funded account. These cards are often issued by finance companies rather than mainstream commercial banks, and they are marketed as gift cards or for use in internet transactions by young people, those who do not have a bank account or, in some cases, to allow the customer to remain anonymous (for example, on adult websites). This is a rapidly growing sector in North America and Europe, where it is often linked with some form of loyalty scheme (which encourages the 'unbanked' to provide sufficient data and transactions to become customers).

15.1.5 Customer cards

Some banks offer customers cards to identify themselves in branches or at ATMs; these cards may allow access to specific areas or services, or they may simply help the cashier to understand the needs of the person they are serving.

Nearly all retail banks now offer some form of e-banking service, both as a convenience to their customers and to reduce their costs for branch-based and even ATM transactions. Take-up of these services is high and closely correlated with home internet usage in each country [1].

Various means are used to identify the customer to the e-banking service: a simple ID and static password are no longer considered adequate, since there are several ways to capture and replay such sign-on procedures. Among the alternatives, most of the more secure options involve a card in some form: to store a biometric or PIN, to generate a one-time password, or to encrypt or sign messages exchanged with the bank. German banks have been at the forefront of moves to ensure that all transactions are digitally signed using a secret shared by the bank and account-holder.

15.1.6 Commercial cards

For higher-value and business transactions there is a range of card-based products available: as well as business credit cards, banks now offer purchasing cards, primarily for online use, which limit the goods and values that each staff member can purchase and enforce a purchasing discipline, as well as facilitating an information flow that includes full invoices with tax amounts.

Smart cards are also used to authorise transmission of payment files (direct debits and credits) to automated clearing houses, and even to encrypt messages for inter-bank transfers. The flexibility of the smart card is such that it encompasses the full range from transactions of a few cents up to very large sums.

15.2 Micropayments and cash displacement

All the card applications described above are intended to help customers carry out transactions in much the same way as they have always done. With the advent of smart cards and e-payments, however, the banking industry has a new agenda: to bring many more payments within the scope of their systems. There are two benefits to banks in doing this: they gain additional information about their customers that can help in customer relationship management, and can start to reduce the cost of cash handling, which, according to the European Payments Council, represents some 0.5% of GDP in Europe and hence over 1% of retail turnover; currently the banking industry absorbs this cost.

They are seeking to do this in two ways:

- By converting to cards and electronic form payments that have traditionally been carried out using cash, such as newspaper and coffee purchases, typically for values less than €15;
- By enabling new transaction types that would otherwise not take place (e.g., a charge for reading an item in an online newspaper, or a charity donation from a cinema seat).

In general, the approach is to reduce the unit costs of such transactions by having most of them take place offline (card-to-terminal or card-to-card) and then settling an aggregate amount. The card itself then plays an important rôle in the security scheme, by acting as the issuer's agent in authorising offline transactions.

15.3 Threats and attacks

The main threats to the integrity of a bank's card systems are those that allow an impostor to carry out transactions, posing either as a valid customer or as a staff member. There are many variants on this, including:

- Use of lost and stolen cards: it is very rare that signatures are checked thoroughly, and indeed if the threshold for a signature check is set high enough to deter forgers then it will turn away a much higher number of valid customers. As shown in Chapter 5, even a four-digit PIN offers a much more effective prevention method for this type of fraud.

- Counterfeit cards: modern printing methods allow a fraudster to create a copy of a valid card good enough to pass any visual inspection, while magnetic stripes can be copied, or even modified, using very cheap equipment. Bank fraud detection systems will detect a certain proportion of counterfeit transactions, but the level of fraud from this source on magnetic stripe cards is still unacceptably high. This is why the international card schemes, and many national schemes, have encouraged or mandated the use of smart cards using the EMV standards (see below), usually by passing the liability for any fraud to the party (card issuer or acquirer) that has not used EMV.

- Repudiation: retailers can invent false transactions, or customers may dispute valid transactions. Again, the smart card offers a digital certificate that can be used to prove whether or not the transaction took place and the key facts about it (time, amount, etc.).

- Direct attacks on cards and keys: several researchers have pointed out theoretical vulnerabilities in cards and in the techniques used to protect them; no cryptographic system is uncrackable and there is always a minute chance that the fraudster's first guess at a key, however long, is the right one. However, card manufacturers and software companies are aware of all current threats and are able to keep ahead of all but the most advanced laboratories. Even where a vulnerability in a current system can be detected, the effort needed to exploit that vulnerability is likely to be excessive and the other controls in place will quickly detect the attempt. However, this does mean that banks must continue to use best current technology and cannot afford to ignore any new attack.

- System attacks: as with all computer systems, fraudsters with some inside access can exploit back-door attacks or missing controls – this is a factor in a very high proportion of successful frauds. Banks' internal systems for authentication are increasingly integrated with the external controls, and make similar use of smart cards.

15.4   Standards

15.4.1   EMV

In the mid 1990s, Europay, MasterCard and Visa developed a common standard for credit and debit-card payment using chip cards. This set of specifications is now maintained by EMVCo, a body set up for that purpose; it currently addresses only contact cards, and does not cover electronic purse or contactless transactions (although the latter are now on EMVCo's work list).

The EMV standard defines primarily a card-to-terminal interface; it does not specify how the card or terminal must work internally, although it does impose a transaction flow and some constraints on the parameters in the card and terminal. It is backed up by scheme-specific rules and requirements, and will normally be referenced by a supplier or equipment purchaser as part of a more detailed specification covering other hardware and software requirements, prompt sequences and card or terminal management.

The EMV standard is conceptually divided into two 'levels': level 1 comprises the electrical and protocol layers, which draw heavily on ISO 7816. Level 2 covers the software requirements, including message and data structures, transaction flows and security, and user interfaces. Any card or terminal that processes EMV transactions must be certified by an independent laboratory as meeting both level 1 and level 2 requirements and, in addition, most card schemes and banks have their own test sequences.

The effect of this is that bank payment smart cards and transactions are indeed highly standardised round the world. However, the cost of cards and terminals, systems changes and management, and the effort of certification all combine to make EMV migration a large-scale task that few banks, and still fewer retailers, will undertake until they are faced with the necessity to do so. In the next case study, I show how EMV has only gradually been adopted round Europe.

Case Study E – EMV deployment in Europe

Background
Although the EMV standard was drawn up by Europay, MasterCard and Visa between 1994 and 1996, its use has not been formally mandated by scheme rules; individual banks were free to make their own decisions as to whether and when to issue EMV cards or replace existing terminals. (The schemes did mandate that new terminals in Europe must be EMV capable).

In many countries, banks felt that the benefits would only accrue if the move to EMV was co-ordinated at a national level, while customers could be confused if each bank implemented the standard in a different way. In those countries, responsibility for co-ordinating implementation was passed to a national inter-bank organisation; in the case of the United Kingdom, a specific body was created to manage the roll-out, as will be described in Case study P in Chapter 20.

Those countries that have an independent national debit card scheme (the majority of countries in Europe) also had to make decisions as to whether to migrate the debit card scheme; transaction volumes on such schemes are often several times higher than on credit cards or international debit cards in the respective country. This decision would be in the hands of the scheme operator.

Business requirement
Although the structure and scope of the project differed from country to country, the business requirement was similar in each case. Those countries that already experienced significant fraud, either as issuers or in outlets in the country, were seeking to reduce that fraud, while the remaining countries wanted to ensure that fraud did not migrate to their countries.

The European Commission also intervened, promoting its vision of a Single Euro Payment Area, in which national borders play no rôle and payment instruments operate in the same way in each country. This helped to drive the development of a set of chip-based card payment products that could meet this requirement.

Solution

Most European banks have opted for relatively simple single-application cards, using native operating systems, for their initial cycle of EMV card issuance. Most payment cards are issued on a two, three or five year cycle and this gives the issuer the option to use more powerful cards on renewal.

Many of these cards, even if they only have one EMV application, support multiple datasets, allowing the card to have different characteristics when used domestically or overseas, and in a very few cases also to support both debit and credit functions on the same card (subject to branding issues). Many cards can also support user authentication using the three-domain model.

In Germany, Finland and some central European countries, most banks have also included a digital signature application that addresses the wider need for authentication in e-commerce, while as we will see in other case studies, some banks have also added loyalty applications to their cards.

Problem encountered

The main problem experienced by individual banks was the need to make a business case, in competition with other projects. When the business case was made at the national level, the rationale for migration was much clearer, but remained a long-term project. The international card schemes encouraged early migration through the use of liability shift (moving the liability for fraud to the party that was not EMV capable) as well as differential pricing and interchange.

Outcome

Progress has been much faster for the international card schemes than for national debit schemes. When the liability shift came into effect in Europe in January 2005, nearly all merchant acquirers in Europe had upgraded their central systems, and were able to accept EMV transactions from capable terminals. However, the actual number of terminals was only around 1.4 million out of 4.6 million in Europe.

The largest issuers in most countries had also upgraded their issuing systems, or were using bureaux that were EMV capable. But smaller issuers remained slow off the mark, and in many countries the implementation had gone no further than some small pilots. At the end of 2004, only 187 million cards had been issued, out of over 500 million.

There were two clear groups of countries for both issuing and acquiring. The UK and France, both of which had national programmes with clear goals, were ahead on both fronts. In Spain, Portugal and Italy some terminals had been placed in order to avoid fraud migration in tourist areas, but elsewhere terminal penetration

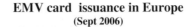

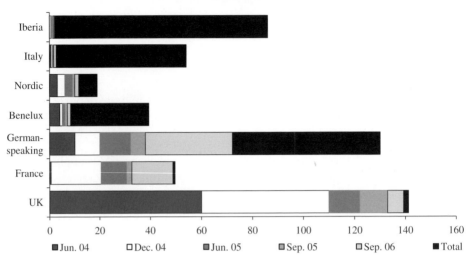

Figure E.1 EMV card issuance in Europe – September 2006 (source: European Payments Council; European Card Review)

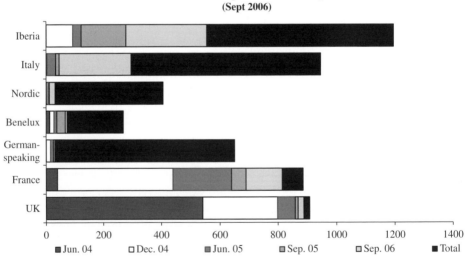

Figure E.2 EMV terminal deployment in Europe – September 2006 (source: European Payments Council; European Card Review)

remained low. On the issuing side, Austria, Belgium, Denmark and Germany were all rolling out cards at normal replacement rates.

By the end of 2006 more progress has been made (see Figures E.1 and E.2) and the first SEPA-compliant products have been launched. Full EMV compliance throughout Europe remains, however, a distant goal.

The move to chip represents an important infrastructure upgrade that will pave the way for many security, efficiency and usability improvements over the coming years. It should be possible to address several transaction types that have been difficult to handle using legacy systems (such as hotel check-in and check-out, unattended petrol sales or discount sales). However, having a global standard also imposes a long gestation time for any changes and it is only a matter of time before variants and departures from the standard become widespread.

The EMV specification has some characteristics that are important when combining payment with other applications in a multi-application card:

- One EMV application can handle several parameter sets, each defining a card type: for example a credit card, international debit card and domestic ATM function;
- EMV uses public-key cryptography for authentication of the card to the terminal (so a terminal need only have a small number of scheme public keys loaded) but symmetric encryption from card to issuer (the card carries secret keys that the terminal does not need to know);
- The specification does not assume that the communications channel from terminal to issuer is confidential: even if data are captured, they cannot be changed, replayed or re-used;
- Although a terminal can process both EMV and non-EMV cards, there are specific rules as to how the initial application selection takes place and it is much easier if the context of the transaction makes it clear, before the card is inserted, what processing will be adopted.

Cards that are used only or primarily for online transactions usually use a form of card authentication known as static data authentication (SDA); this consists of a certificate signed by the card issuer, which is sent to the terminal for verification during a normal transaction. Since SDA certificates can be copied along with other card data, this leaves a loophole for a counterfeiter who can gain access to valid blank cards and who can be sure that any given transaction will not be authorised online (EMV includes random online checks to avoid this). Issuers are gradually moving to the more secure dynamic data authentication (DDA), which uses the card's private key in every transaction and thereby closes this loophole; this requires the use of a higher-specification card including a crypto-processor.

15.4.2 Contactless cards

Following various pilot projects, mainly in the petrol retail industry, MasterCard published in 2002 the specification for its *PayPass*^{TM} contactless interface. PayPass uses a specific implementation of the ISO 14443B electrical and protocol standards, while for its messages and data it allows either of two options:

- A 'magnetic stripe' mode that transmits the equivalent of the magnetic stripe data, together with a transaction counter; this option is used in North America and allows a contactless interface to be retrofitted to a legacy magnetic stripe terminal.

- An extension to M/Chip^{TM}, the MasterCard version of the EMV standard; this option is used in Europe and Asia and extends the capability of a chip-reading terminal.

In 2005, Visa licensed the PayPass implementation of ISO 14443 for its Wave^{TM} contactless programme, so that terminals can easily be adapted to accept both types of card, while American Express also uses ISO 14443 for its ExpressPay cards. This means that at the electrical and protocol level, all three schemes are broadly compatible, but there are still significant differences in the software and data used for each. It is also important to note that none of them is compatible with the MiFare^{TM} or FeliCa^{TM} standards widely used in the transportation industry, although terminals can be designed to meet both sets of requirements.

15.4.3　Electronic purses

Most of the first-generation electronic purses used proprietary (national) specifications; even Visa's VisaCash^{TM} programme included multiple standards that reflected several national specifications. The Mondex system, later acquired by MasterCard, was an exception: it was implemented in several countries in a more or less identical format. The Belgian Proton system was also implemented in some other countries.

In 1999, the European Committee for Banking Standards published the Common Electronic Purse Specifications. Rather than seeking to impose a common standard on all existing purse schemes, CEPS comprises a set of common functions that can sit alongside an existing purse scheme to provide interoperability, or can be used to form a purse scheme on their own. However, the drawbacks of electronic purses referred to above were becoming clear at the time CEPS was published and so this specification has been little used.

Stored-value programmes remain common in closed-circuit applications, such as campus schemes and transportation, and as banks take a greater interest in multi-application cards, it is likely that there will again be a move towards open schemes. Pre-authorised debit (which can be implemented on EMV-compatible cards and terminals) does not meet all the requirements of these schemes and it is likely that some form of electronic purse standard (probably with a contactless interface) will re-emerge in the coming years.

15.4.4　Token authentication

Of the various methods currently in use and on offer to provide secure access to internet banking and e-commerce transactions, possibly the most dependable is the use of a token combined with a biometric, PIN or other password. This combination of 'something you have' with 'something you know' and 'something you are' offers the greatest security.

One option is to have a smart-card reader attached to the user's PC (the readers themselves are very low-cost items and could be included in the internet banking package charge). The card would then communicate directly with the issuer's system to authenticate the user and authorise transactions. From a security point of view, however, it is much easier to define and manage the risks if the card reader is offline: MasterCard's Cardholder Authentication Program (CAP) allows an EMV card to be used in this way. The card is inserted into a reader such as that shown in Figure 7.4, the card-holder enters his or her PIN, and the display shows a one-time password that the card-holder can type into the website. This allows the card issuer to authenticate the user on entry to the internet banking system and to authorise any transactions (with a second password, if necessary, for any sensitive transactions).

Variations on this structure allow the card to respond to a 'challenge' issued by the website, or to sign details of the transaction in the same way as for an EMV transaction.

In the future it would be very easy to allow an EMV card to carry a fingerprint template or other biometric that could be used in this environment – many PCs already have fingerprint readers.

15.4.5 Others

The banking industry makes use of a large number of specifications and standards covering security, message exchange, etc. Some of these are international while others have a purely national scope. Some (such as EMV) require a very detailed and lengthy certification process, while others are handled by supplier self-certification or testing by the bank. Any supplier or industry partner wanting to work with a bank must be aware of the existence of these standards and be prepared to work within them.

15.5 E-payment and m-payment

To address the problems of user authentication, confidentiality and non-repudiation inherent in any remote payment system, the card schemes have adopted a *three-domain model* (see Figure 15.2) that allocates responsibility for user authentication to the card issuer, for merchant authentication to the acquirer and for interoperability to the card scheme.

For e-commerce, the user authentication requirements can be met using the token authentication methods described above. M-commerce (using a mobile phone network rather than the internet) poses some additional problems, as we saw in the previous chapter. Solutions have been implemented using two-slot telephones, 'wallet' systems, where the user's card details are stored at the network host (reverting to the legacy model of magnetic stripe payments, since there is no authentication of the card details and no data are written back to the card after the transaction), and multi-application cards, as in the Moneta example cited in Chapter 14.

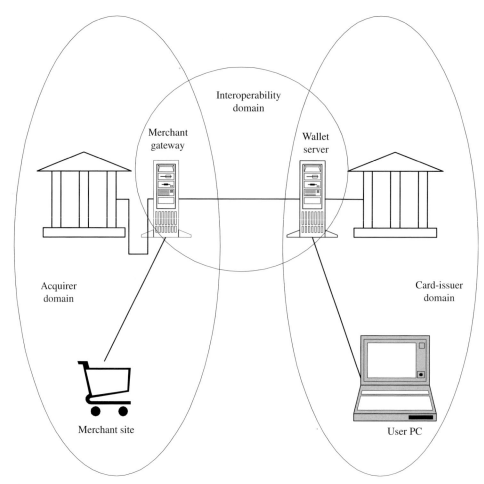

Figure 15.2 Three-domain model for internet payments

15.6   Loyalty

It is nearly always cheaper to retain a customer than to recruit a new one; all service providers, including banks and retailers, therefore, seek to build loyalty in their customers. There are many ways to do this, but two methods that are often employed both make use of cards.

One is to understand in every detail the customers' buying behaviour, to build a picture of customer needs and to be able to respond quickly to changes in those needs. The other is by offering rewards and discounts for high and regular spending. It is relatively easy to design a loyalty scheme using cards, although implementing the scheme (and in particular delivering the rewards) can prove more tricky, and in fact there are many possible variants. Smart cards are of greatest value where the rewards are relatively high (since they prevent most frauds) or where points may be awarded and rewards offered in many locations and businesses (since the points and transaction details may be stored on the card).

Loyalty is, therefore, often quoted as an application that can add value to a payment card. However, the question is always: loyalty to whom? To build its own loyal customers, banks have all the data they require on every transaction in a central database – this does not require a card. A partnership with a single retailer or service provider means that the bank is taking a competitive position in that retail sector, which is likely to draw opposition from other competing businesses. Schemes that allow the bank to partner with a relatively wide range of retailers are, therefore, the most likely to succeed, but they may be difficult and costly to administer.

In some countries, however, a loyalty scheme is seen as a necessary feature of a credit card (they are much less commonly used with debit cards). In these cases the issuer must study carefully the costs and benefits the programme will bring. Although most card suppliers are able to offer an off-the-shelf loyalty application, a programme that has a good strategic fit with the environment, customer base and the issuer's objectives is much more likely to succeed than even a very comprehensive scheme that does not meet these criteria.

In 2001, Target Corporation, a major US retailer, decided to extend its loyalty offering by including a smart-card-based coupon scheme which offers rewards directly relevant to customers' past buying behaviour. This was offered as an additional application on its Visa credit card, which was upgraded to EMV. All card-holders received a smart-card reader, which enabled them to access their account and download coupons to their cards from home. In March 2004, Target announced that it would phase out the smart-card element of its Target Visa programme. At the time of the announcement, the Target scheme was the largest EMV programme in North America, with some 7 million cards. Although the cost of cards and readers is likely to have been a contributory factor, the prime reason was the lack of take-up by customers of the coupon-based loyalty scheme.

The following case studies, however, provide examples of loyalty schemes that have been considered very successful by the card issuer and have generated significant additional revenue.

Case study F – Mashreqbank WOW! card

by Wong Wan Ling, Group Product Marketing Director, Welcome Real-time

Organisation

In January 2004, Mashreqbank, one of the oldest banks in the United Arab Emirates and the leading merchant acquirer with around half of the UAE acquiring market, launched the WOW! card, which is the first chip card in the Middle East that offers customers value-added features on top of payment.

Mashreqbank used the same infrastructure to launch a co-branded card with Virgin and MasterCard, offering invitations to concerts and movie previews, in-store promotions and other special benefits.

Business requirement

Mashreqbank's objective was to persuade card-holders and merchants to view payment as something new, exciting and different, so that they would focus on the added value provided by Mashreqbank, rather than the cost of Mashreqbank's payment service. Another objective was to enlist merchants' enthusiasm for the value-added features so that they, in turn, created excitement around Mashreqbank's WOW! card in their stores.

Most merchants already offer their customers paper-based incentives, such as coupons distributed in the local paper. Rather than distribute the same incentives to everyone, Mashreqbank merchants can now use the bank's service to deliver their highest value promotions to their best customers. They can also manage soft benefits like welcome gifts, surprise gifts, VIP access, or special upgrades, more effectively.

Solution

Mashreqbank deployed Welcome's XLS payment software for its WOW! card implementation. Each card carries an XLS 'application' or data file, and the host application (which runs on Mashreqbank's merchant acquiring host system) delivers merchant-specific, customised, and targeted promotions to card-holders, based on their purchasing behaviour.

The XLS software is integrated into Mashreqbank's payment infrastructure and the promotional marketing payment features are offered to merchants as part of the payment services offered by the bank.

Merchants are encouraged to run their own specific promotions and incentive programmes targeting customers who are utilising the WOW! card. The value of the incentives and rewards to card-holders varies, with high value rewards going to higher spenders. The promotions are real-time, instant, and entirely automated, with little or no additional effort at the point of payment.

Outcome

Mashreqbank's new chip infrastructure produced substantial benefits in its first year (Figure F.1):

The card-holder base grew by 65%, from 85 000 cards at the end of 2003 to 140 000 cards at the end of 2004, far exceeding the overall UAE market's 15% growth. By June 2006, the card base had grown to 300 000.

Mashreqbank's merchant network grew by 25%, from 6000 merchants at the end of 2003 to 7500 merchants at the end of 2004, in an already mature and saturated acquiring market. Mashreqbank now represents around half of the UAE acquiring market. This grew further to reach 9200 merchants in June 2006.

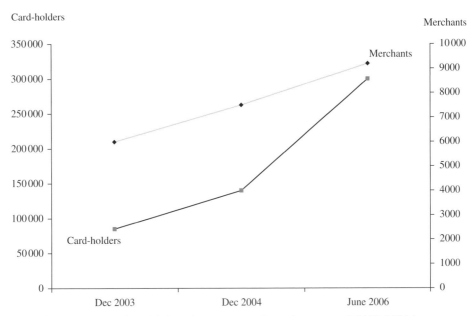

Figure F.1 Mashreqbank WOW! card customer and merchant growth 2003–2006 (source: Mashreqbank)

New card-holders are bigger spenders than typical Mashreqbank customers. Average spending per card, across all Mashreqbank card-holders, new and old, increased 13% and is now equivalent to the average amount spent by customers of all other banks, whereas prior to launching WOW!, the average spend per Mashreqbank card-holder was almost 10% below the average of all other cards.

Mashreqbank card-holders are shopping more often at Mashreqbank merchants, yielding marketing reinforcement as well as savings on interchange fees. Spending using Mashreqbank cards at Mashreqbank merchants grew by 23%, almost four times the growth experienced by other banks in the UAE.

Since late 2005, Mashreqbank has also issued chip-based debit cards that enjoy the same merchant promotions as WOW! cards. Mashreqbank sees debit-card payment as an opportunity to improve the profitability of retail accounts, for example increased use of the debit card may encourage higher average balances, allow the bank to offer bundling with other retail products, or create cross-selling opportunities. But since debit cards are issued free of charge and do not earn interchange, it was previously difficult to fund programmes to encourage spending on the card. Use of the chip and XLS application provides this encouragement, and by June 2006, Mashreqbank had 50 000 chip debit cards in circulation.

Case study G – United Bank Limited

by A. Hamid Farid, United Bank Limited

Organisation

United Bank Limited is one of the largest commercial banks in Pakistan, with a network of over 1000 domestic and 15 overseas branches. In 2003, it was privatised and a complete restructuring of the organisation included the introduction of a Consumer Banking Division, which represented a major change to the image and vision of the bank. Since 2003, UBL has launched several new consumer products, of which the UBL Credit Card is seen as the most innovative.

Business requirement

At the time of the launch of UBL Cards, the challenge was to create differentiation from other large card issuers who were already strong in the market, with substantial customer bases but using conventional magnetic stripe cards.

United Bank Limited decided to differentiate its product from other cards using innovation, and so entered the market immediately with a chip card, representing the next generation of credit cards in Pakistan; this was the first chip-based credit card in the country. To ensure the widespread acceptance of these differentiated cards, UBL also set up a chip-based acquiring and terminal infrastructure (although the card also carries a magnetic stripe, so that it can be used in any merchant location).

Market positioning was also important: the card has been positioned as innovative, and representing fun, value and entertainment. Chip technology represented innovation, but for the other aspects UBL adopted a concept called 'dip the chip'. The card allows the customer to gain instant rewards at the point of sale when making transactions using the card. Customers are unaware of the details of the loyalty programme but know that they will gain instant rewards when using the card. This represented a change from the traditional serious image of a credit card to something that represents fun, excitement and value.

Solution

United Bank Limited adopted Welcome Real-Time technology (XLS V6) for its server system and for the card-loyalty application, with a highly modular and parameter-driven software application from VeriFone (Softpay) for the terminals. The loyalty programme is based on four sets of parameters representing recency, frequency, monetary value and accumulated points; these can be used either individually or in combination.

Implementation

Despite the fact that the product introduced was both innovative and useful, UBL initially encountered several implementation problems. The first was that of complexity: the new product was not properly understood by (or indeed explained to) the merchants and customers, for whom it was a totally new concept in this part of the world. In the initial stages, the scope was too broad and did not focus on high-frequency-purchase points; the booklet included a very large number of categories and offers, which made it complicated, lengthy and confusing. As the schemes were funded by the merchants, many offers were less attractive than was needed, while UBL did not have sufficient control of the proposition.

As a result, UBL changed to a more focused strategy. Since customers were making more use of the scheme in restaurants and department stores, efforts were concentrated on these two categories. And the repositioned proposition was funded by UBL Cards itself, which meant that offers at loyalty merchants nationwide were standardised.

United Bank Limited had some difficulty selling the proposition to merchants when it was completely new and the benefits were unproven. At the time of the launch the system required a 'double dip'; this was later converted to a single dip. Communication issues resulted in a higher-than-expected rate of decline, while batch settlement and payment-timing issues were also frequently encountered. However, with time and experience, most of these issues have been resolved and the processes and systems now work well and effectively.

Future development

United Bank Limited's future strategies include finding new profitable avenues where its loyalty programmes can be run. The frequency of credit-card transactions at fuel stations is high and this represents an immediate target for 'dip the chip' loyalty programmes. To reduce the cost of the instant rewards, negotiations with consumer goods manufacturers are under way to allow the latter to contribute marketing expenditure in return for increased visibility at the point of sale.

15.7   Co-branding

Many banks issue co-branded cards, sometimes with an airline or other service provider, but most often with clubs and other affiliation organisations. Co-branded cards offer the greatest opportunities for banks seeking to issue multi-application cards: many of the likely applications are relatively simple and there are few commercial or liability implications.

For example, where the co-branding partner is a club, the card may offer proof of membership, a record of status or activities, and even a stored-value function for use in club shops and bars. A card co-branded with an insurance company or savings scheme could carry a record of contributions.

The most complex, and yet potentially the most rewarding, co-operations are those between banks and other mainstream card issuers, such as telephone companies, governments or transportation companies (including airlines). These will each be considered in their respective chapters; however there is a common issue: the bank not only expects to perform the rôle of the card issuer in all these cases, but is, in many cases, required to do so by law or by the rules of the card scheme that it belongs to, or both. It is extremely difficult for the bank to take a junior, or even an equal, rôle in any co-branding partnership. This can raise organisational problems, as I will discuss in Chapter 19.

One of the most tricky problems for a bank is the need to make 'know your customer' (KYC) checks when a new account or relationship is opened. To meet international standards in money-laundering prevention, these processes often involve checks of physical documents. This requires a costly manual process and delays the signing-up process; it can be a significant barrier to a bank wanting to issue cards in conjunction with another organisation, even if that organisation already has a complete mailing list with all the necessary details. This may often feel like unnecessary bureaucracy to the partner, but is required in all countries that sign up to the international anti-money-laundering treaties.

15.8   The future of multi-application cards in banking

Banks see themselves as special among card issuers: certainly they are one of the few sectors for whom cards are, and have been for many years, a core business. Their security requirements are well-defined (but not necessarily higher than those of other issuers), while bank card schemes played major rôles in shaping both GlobalPlatform and Multos.

Banks have a strong interest in adding value to their customer relationships, and see multi-application cards as one way to achieve that – they are likely to develop a range of products, initially by mixing and matching the applications described above. However, structural and organisational barriers will probably restrict the extent to which banks themselves work with other sectors or provide non-financial applications. It is possible that non-bank payment service providers, which now provide an ever-increasing part of payment systems functionality, may be better placed to bridge this gap between banking, payment and other application sectors.

15.9   Reference

[1] Meyer, T. *What We Learn from the Differences in Europe*. Deutsche Bank research paper February 2006

16 Transportation

Of the many skills needed to operate a bus company, airline or train service, ticketing and card issuance would not normally rank highly on the list. But transport operators are increasingly turning to cards to protect their revenue and to make passengers' journeys smoother.

16.1 Existing public-transport card schemes

Most existing public-transport schemes in cities, towns and rural areas are based on trains and buses for long-distance travel, combined with buses and trams for local journeys. These are often linked together under the auspices of a local transport authority or consortium, and may offer some form of common ticketing system and fare structure.

16.1.1 Revenue management

The business case for existing public-transport operators to convert their ticketing schemes to smart cards is often very strong, and is based on improving revenue management and reducing operating costs.

Each operator, particularly in a group or consortium, wants to ensure that it receives the revenue to which it is entitled and this demands a shared pool of information about passenger journeys as well as costs. With older forms of ticketing (paper or magnetic stripe tickets) this could only be achieved with great difficulty, if at all; the cost of collecting the data was very high. It was also difficult to check the cash collected by on-board staff. With a smart-card ticket, the card forms part of the data collection system and a full record of all transactions can be collected. This enables revenue to be shared more accurately, and thus encourages common ticketing schemes.

Removing cash handling from the bus or train improves security and reduces cash 'leakage'; checking is easier and more automatic. Details of each journey can be stored on the card as well as in the terminal, and this opens up new areas such as interoperable journeys and price capping. A card-based system also greatly helps in planning and managing service growth; typical journeys and use patterns can be traced and used to improve the service.

16.1.2 Speed and convenience

For customers, wider ticket sharing is convenient, since they have to buy fewer tickets; in particular, journeys that involve several legs or different means of transport are made much easier, since they waste less time queuing for tickets.

By comparison with magnetic stripe cards, contactless smart cards allow a faster passenger flow through turnstiles or when boarding a bus – the benchmark for a completed transaction is 200–300 ms. If the alternative is a visual check, the smart card is much more accurate and requires less manpower as well as taking less time. The speed advantage is not shared by contact cards, which is the main reason why such cards are not generally used in public transport.

16.1.3 Operating costs

Another reason for preferring contactless cards is their low maintenance costs: readers are often in public places and are subject to wear, vandalism and environmental damage. With a contactless reader, there are no moving parts and the unit can be fully sealed (see Figure 16.1).

The card can store significant volumes of data: even the simplest schemes use a 1 kB card, sufficient to store ten or more journeys, and those data can be re-written as often as required. The card material is usually sufficiently strong for cards to last several years, particularly where contactless cards are used. Most current cards (e.g., MiFare™ 1 K or FeliCa™) are very cheap (well under $1), but even higher-value cards can be justified if the life of the card is several years.

One area for caution is the possible support costs for a card-based scheme: public-transport customers are often not very careful with their cards, which can be lost or damaged. The cost of replacement cards is higher if there are personal data (and particularly a photograph) on the card, but the frequency of loss may be much lower. And a call centre is likely to be needed to handle support questions such as, 'My card was not accepted at a gate today – what must I do?'

16.1.4 Interoperability

Many public-transport customers use several forms of transport or transport operators for their regular journeys, or use different operators at different times: for example, train followed by tram for the journey to work, but bus if travelling in the evening. For these people, it is a significant advantage if tickets bought in one location can be used in another – this also allows 'through ticketing', where one ticket covers all the forms of transport needed on a particular journey.

Figure 16.1 Oyster card reader (© Transport for London 2005)

Operators disagree about the optimum scope for interoperable ticketing – is it an extended metropolitan area, a region or a country? Is it important to link long-distance and local travel? The answers to these questions depend on the population and travel patterns in any given area; however, there is little doubt that using a form of ticket or token that allows interoperability, and having fare structures that can accommodate it, gives the flexibility that allows operators to design schemes appropriate to their area or business requirement. Later in this chapter, we see how this is being addressed in Europe by the ITSO and Calypso standards.

Case study H – Lisboa Viva and 7 Colinas

Organisation

The Lisboa Viva and 7 Colinas cards are issued on behalf of 21 transport operators that come together under the umbrella of OTLIS (Lisbon transport operators' consortium). The network includes the city bus and tram operator CARRIS, the 'Metro Lisboa' underground system, as well as ferry, train and private suburban bus operators. In addition to the 2.5 million inhabitants of the city, it has to serve 7 million tourists each year.

Business requirement

In 1998, the Metro Lisboa recognised the need to replace its ageing ticketing and access control system with one that would support conversion to the euro and other future developments. Lisbon already had a multi-modal ticketing system, based on magnetic stripe cards and with a very complex fare structure (over 8000 ticket parameter combinations). The operators wanted to retain the integrated tariff, but to improve control and the fairness of the revenue split; they also wanted to build a database that would help with planning the development of the network.

Solution

Lisbon was used as one of the test beds for the Calypso system, and benefited from support from the European Commission's e-Europe project.

The core of the system is the multi-operator Integrated Intermodal Transport System (SIIT), a set of database applications developed by Link Consulting. This provides management of ticketing products, cards and personalisation, loading and reloading, validation transactions, and also global security over all the card transactions – see Figure H.1. The SIIT system is linked through internet connections to the individual operators' consoles and ticketing systems; middleware provides the link to the database so that all applications are developed in a standard, technology-independent way and can accommodate new vendors' systems or technologies with little additional development.

At the terminal level, OTLIS, through its implementation partner Link Consulting, decided to build a highly generic and open system, using a formal embedded software interoperability framework that avoids dependency on any one hardware supplier.

Initially, in 2001, the operators selected cards from French manufacturer ASK, in two formats:

- 'Lisboa Viva' cards are standard Calypso dual-interface cards using a fixed file format (with master files, dedicated files and elementary files of many types, such

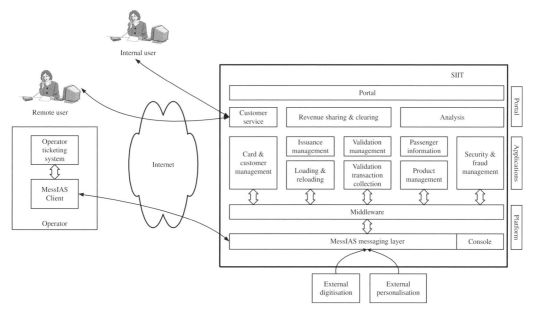

Figure H.1 SIIT system diagram for Lisboa Viva card (source: Link Consulting)

as cyclic, transparent binary and counters), DESX encryption and DESX-based mutual authentication between card and SAM. The card body is PVC and is printed with the user's photograph and name, using a desktop printer at one of around 30 stations and customer service points. Currently, up to four simultaneous combined-ticket products can be loaded onto each card.

- '7 Colinas' cards are paper 'C-tickets', with contactless interface only. For commercial reasons, they can only be used with the simpler product types (e.g., daily, n-day or 30-day passes) but can be reloaded up to 127 times. They are not personalised, and are issued either from ticket machines (dispensed from a roll) or in bulk (e.g., through tour companies or conference organisers).

Based on the embedded software interoperability framework, new card suppliers, with new card products, were introduced in 2004. Further developments currently being undertaken will allow more product variations, but still without disruption to the existing multi-vendor system.

Problems encountered

Metro Lisboa recognised that this was a complex multidisciplinary project with a short timescale for delivery. Strong leadership and hands-on involvement of senior management were both essential.

Lisbon's geography and the many operators involved have resulted in a very complex fare structure, with several thousand ticket products. The underlying Calypso architecture was able to accommodate this, while SIIT was developed using open standards wherever they existed, and with a formal data-model structure, to ensure that no limitations would be imposed. The operators intend to include a 'prepaid' or 'pay-as-you-go' facility; the card and data structure will both permit this but the complexity of the fare structure means that the calculations at each entry and exit point would be impractical and so greater effort is being expended to try to reduce this complexity.

The dispensing of the paper '7 Colinas' cards in ticket machines caused some problems at the outset. Because the antenna is relatively close to the card edge, small misalignments can result in its being damaged or even cut by the dispensing mechanism. Software changes were needed to ensure more accurate and consistent cutting.

Initially, all tickets were issued by individual transport operators (albeit using the central database and centrally designed facilities). The introduction of ATM and internet reloading generated a new requirement, for central customer service facilities not tied to any individual operator. In practice, these have been provided up to now by the two main operators: CARRIS and Metro Lisboa. It also resulted in the need for a central clearing function, rather than the manual accounting system used previously; however SIIT had been designed as a service centre with central clearing in mind, and this was implemented in 2005 and 2006.

Outcome

Development work started in September 2000 and was complete a year later in time for the introduction of the euro. The functionality has since been extended to include internet reloading (using a PC-attached contact reader), ATM reloading (this facility is now available at any ATM on the SIBS inter-bank network) and, also, a Payshop (small-retailers) POS network.

In 2005 the operators launched a card for tourists: the LisboaCard, which also includes public-transport entitlement and is based on the same principles and software platform; 100 000 of these cards are now issued each year.

16.2   Non-transit usage

Several public-transport smart-card scheme operators have expressed interest in extending the usage of their cards to the purchase of non-transit goods and services. This may be for social reasons: many public-transport users do not have bank accounts, but would like the convenience of using a card. The motive may be purely commercial: there is potential revenue in facilitating some payment categories (although in the previous chapter I showed that this is not always profitable). Or it may be to encourage passengers to use the transit card by giving the card more functionality.

Since the card is likely to be contactless, the best locations and transaction types are those where speed and convenience are important. The amounts should not be too large: there is probably no signature or PIN entry, so there is a risk that the card may be stolen or counterfeit. And if there is a mental link between the purchase and the journey – a coffee to drink on the way, for example – then it is more effective in encouraging use in transit as well as shops.

Case study I – Kaohsiung City Government 'TaiwanMoney' card

Organisation

In October 2005, Kaohsiung, the second largest city in Taiwan, launched the Smart Transport Card Project as part of its initiative to transform Kaohsiung into an *e-City* through the deployment of next-generation technologies. Cathay United Bank and E.Sun Bank agreed to issue the cards, supported by Mondex Taiwan, which operates the clearing and settlement service, with Acer acting as the systems integrator.

Business requirement

The underlying objective for the Kaohsiung City Government (KCG) was to accelerate the city's economic development.

The Kaohsiung City Government wanted to introduce a single smart card, with a single balance, to enable both retail and transport contactless payments, thereby improving the efficiency and convenience of the city's transport system. In doing so, KCG was keen to avoid the hidden costs of card issuance (production, distribution, customer support), the management of the scheme (risk, clearing, settlement) and legislation (for example, obtaining banking licences).

The solution covers seven cities in southern Taiwan, with a total population of around 6.5 million, and is managed by KCG. The cards are used in buses and on ferries in Kaohsiung and the surrounding district, and also in shops using standard EMV terminals (usually equipped with a contactless reader).

Solution

Rather than trying to expand a closed transport-card scheme into retail payments, KCG and its partners have opted to extend an open retail-payment scheme into their transport system.

Cathay United Bank and E.Sun Bank issue both EMV credit cards with the MasterCard Cash function and anonymous cards that only contain the MasterCard Cash function. This is a true electronic purse, held by Mondex Taiwan and subject to supervision by the Taiwan Banking Bureau; there is no need for card-holders to set up an account at the bank or elsewhere, although they may if they wish link the card to an existing account for top-up purposes.

Both linked and anonymous versions are 32kB Multos dual-interface (contact and contactless) cards and use a standard M/Chip 4 payment application, but with some proprietary data elements that define transport-related data (concessionary ticket types and a journey log) as well as the standard M/Chip contact and contactless memory areas. Some terminals use the ticket types to give special fares or free travel (for example, inhabitants of a small island receive free ferry travel to and from the main island), but most journeys use the standard EMV functions to decrement the balance held on the card. City buses and ferries work with a fixed-fare structure, but on longer-distance buses the passenger must tap in and tap out in order to pay the correct fare; if they do not do this the card is locked and must be reset by an operator.

When the card is used in retail outlets (coffee shops and other low-value outlets) the transaction is handled using a conventional EFT terminal with a contactless reader. These terminals accept both MasterCard Cash and MasterCard PayPass and can also be enabled to accept any contactless payment meeting the MasterCard PayPass standard (which both Visa and JCB have licensed). The validator on the buses and ferries is an attractively designed box roughly the shape of Taiwan, with the TaiwanMoney logo indicating the 'sweet spot' (see Figure I.1).

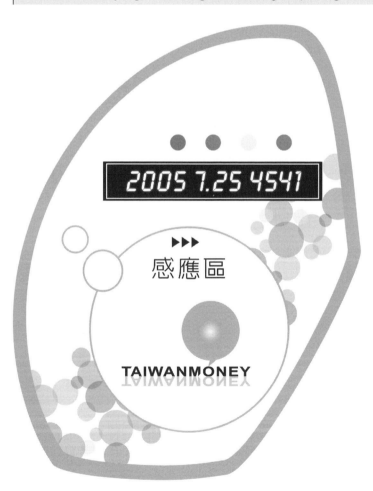

Figure I.1 Bus validator for TaiwanMoney (courtesy of MasterCard Taiwan)

Both terminal types display the remaining balance, and the card can be topped up (by transfer from the credit card account or using cash in the case of the anonymous cards) at ticket offices, branches of the relevant bank and at some of the retail merchants. The 11 transport operators involved are responsible for collecting the transactions from individual bus terminals (using GSM) when the bus reaches its depot; these are concentrated in the operator's network and sent to Mondex Taiwan, which sends the journey information to the transport operator's Head Office and the financial information to the banks. For retail transactions, Mondex Taiwan collects the transactions directly from the terminals.

In operation, transactions are a simple tap-and-go operation. As with a normal *PayPass* transaction, three lights switch on in sequence. Purse transactions are a little slower than most transit purses (such as those using MiFare^{TM} or FeliCa^{TM}) but at less than 600 ms are acceptable when boarding a bus.

Status

A pilot was launched in October 2005, covering city buses and a limited number of retail outlets. The participants wish to extend it to other parts of Taiwan, but the system requirements for these extensions are subject to the business agendas of the relevant local governments as well as other transportation companies.

There are plans for the transport operators to implement e-coupon and loyalty programmes offered by banks, to increase card usage and generate new revenue. The Multos platform allows both proprietary and off-the-shelf applications to be added either before or after issuance.

There are several barriers to this extension of usage: I will discuss security and card types at the end of this chapter, but in some countries it may even be illegal for a transport operator to offer a means of payment: this activity is reserved for banks. In Europe and in much of Asia, however, there are specific electronic money provisions in the law to allow non-banks to offer such facilities.

Case study J – EZ-Link/QB

Organisation

EZ-Link Pte Ltd is a subsidiary of the Singapore Land Transport Authority and operates the EZ-Link card, which is the only electronic fare card to pay for travel on buses and mass rapid-transit (MRT) trains in Singapore. Over 8 million EZ-Link cards have been issued and are used for 4 million journeys a day. Citibank is the issuing bank, and is responsible for the value stored on the cards and for ensuring that the system meets the security and regulatory requirements of the Monetary Authority of Singapore.

QB Pte Ltd, a Singapore based company, has an exclusive agreement with EZ-Link to offer payment and loyalty services to non-transit merchants using the card.

Technology

The EZ-Link card is based on Sony FeliCa™ technology to provide rapid and simple payment transactions from a single purse, as well as the loyalty services offered by QB. Readers are very simple 'tap-and-go' devices, with a display but no keypad. Transactions are authorised either online or offline, and uploaded at the end of each day or at regular intervals. Work has also commenced on the possibility of issuing ISO 14443-B cards, with single-trip tickets already issued for use.

Transaction types offered

The EZ-Link card was first issued in 2002, and in 2003 QB started to offer contactless payments in retail outlets, mainly in fast-food outlets, vending machines and convenience stores. This was followed by a charity donation programme, in which card-holders can donate to charities simply by tapping their cards. Other outlets accepting the cards now include supermarkets, cinemas, government departments and many schools. The EZ-Link card is also used as a loyalty card in the leading fast-food merchant in Singapore.

Currently (mid 2006) all road tolls are paid for using a 'cashcard' – an electronic purse card run by a consortium of local banks. The next generation of in-vehicle units for road tolling will also include a contactless card interface, allowing ez-link cards to be used in them.

And card-holders will shortly be able to buy their own contactless reader terminals, for attachment to PCs, which will enable the cards to be used for internet payments and top-ups. The level of internet penetration and literacy in Singapore is very high, and this facility will enable many card-holders who do not have credit cards to make internet purchases.

EZ-Link and QB have also announced a co-operation with Visa that will allow banks to include EZ-Link functions on bank-issued cards. These cards, when issued, will come with either credit or debit functionality and allow each bank customer to hold a single card that combines retail payment and ATM functionality, and that can be used for transportation needs.

16.3   Inter-modal use

The benefits of interoperability, discussed previously, are further increased if the scheme can be extended to other forms of transport that may form part of the same journey.

16.3.1 Trains

For long-distance trains, the benefits of smart-card ticketing are generally less than for local transport: there are advantages in having printed tickets showing more information, and this is also suitable for checking tickets on board the train. Since most long-distance train journeys are one-offs, the cost of the card is high in relation to the benefit gained.

The exception to this is season tickets, particularly for commuters. One of the most widely used transport smart cards is that of East Japan Railways (JR East), which has a high proportion of commuters amongst its passengers. JR East's 'Suica' card, which uses FeliCa^{TM} technology, offers a number of benefits to card-holders, including the ability to hold both a season pass and 'pay-as-you-go' value on the same card. Some seats also contain a reader, which brings on a green light above the seat if the ticket is valid. This reduces the time required for staff to check tickets on a crowded train.

16.3.2 Taxis

Although taxis in many cities are fitted with card readers and radio modems for credit and charge-card acceptance, it is very unusual for them to be able to accept public-transport cards. The main reason for this is not technical (although the taxis would also need a contactless reader) but structural: taxi drivers are usually self-employed, independent businesses, and find few benefits in working within the structure of a public-transport authority that is used to dealing with larger firms.

The exception to this is in China, where transport cards in many cities can also be used in taxis. Here the transport card operator is separate from the bus company and transport authority (although it is usually still owned by the city government). The taxis are fitted with readers by the transport card operator, and in some cases (e.g., Shanghai and Beijing) the city requires all taxis to accept the cards. In other cities, despite having the readers, drivers prefer to receive cash. The aim of most transport card operators in China is, however, to encompass all forms of public transport.

16.3.3 Road tolling

Electronic toll collection (ETC) schemes offer significant benefits to the road operator [1]. The main one is faster throughput at toll gates: fees can be collected from vehicles at speeds limited only by safety around the toll plaza. This, in turn, means fewer toll gates and fewer operators, while the reduction in cash handling offers further savings. In addition to motorways, bridges and tunnels, ETC technology is now being used to facilitate restricted access or congestion charging schemes for city centres [2].

Electronic toll collection systems use a combination of several technologies, such as toll collection at entry or exit, often combined with a vehicle identification system for

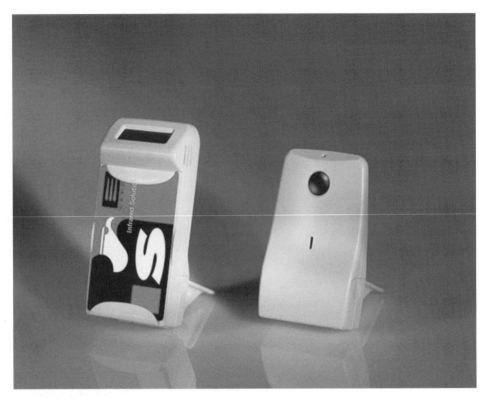

Figure 16.2 On-board unit (OBU) for road tolling (© Efkon AG)

policing. The key to many of these systems is digital short-range communication (DSRC), which uses a 5.8 MHz signal linking transponders at the roadside, in the road or on a gantry with an on-board unit (OBU) in the vehicle (see Figure 16.2).

Digital short-range communication enables the communication between vehicle and central system; the actual payment or entitlement can be through a centrally-held account, but many operators have adopted a smart card that communicates with the on-board unit. In Singapore, for example, this is the contact-based electronic purse 'CashCard', while in the Italian motorways operated by Autostrade s.p.a., the 'Telepass' system uses a contactless card that sits within the on-board unit, in both cases on the dashboard or windscreen. In some other cases, the on-board unit is fitted to the underside of the car; this is more suitable where the transponder is buried in the road.

Although the decision to open the barrier or turn the traffic light green is usually based on an offline transaction, the actual account is likely to be held centrally. Where there is no barrier, drivers must be made aware of the amount remaining in their account or card, and this will often require a display on the on-board unit. One issue with these schemes is: how to deal with non-payment? There must be a vehicle identification system for photographing number plates, linked to the national vehicle registration system (see Figure 16.3).

Figure 16.3 Gantry for road tolling and number-plate recognition (© Efkon AG)

The new generation of contactless cards also permits the use of a standard card where the user slows down and touches the reader. Currently, ISO 14443 cards can be used, although NFC has also been trialled. This allows true multi-modal transport, as we can then use the same card for trains, buses, parking and toll collection.

More sophisticated schemes use global positioning systems (GPS) to track the location of vehicles in the network: this allows complex schemes for charging by location and time as well as distance. They can also be integrated with other vehicle and driver monitoring data: for example, recording the driver's hours and speed, or the temperature in a refrigerated lorry. These data can be stored on a smart card and subsequently uploaded to a central system. Such systems exist already (for example in Switzerland and Germany), but their use is limited by the lack of standardisation across all the countries where the lorries that are the main users of such systems may travel.

16.3.4 Parking

There are various ways in which smart cards can be used in connection with parking. The simplest is to use the card as an access control token, for example in staff car parks or for season-ticket holders: presenting a valid card allows access to the car park.

Figure 16.4 Personal parking meter (courtesy of DXP France)

For pay-as-you-go parking, users can buy tickets or pay on exit using an electronic-purse card (this can be fully offline). For higher values, a credit or debit card may be used, but this must usually be an online transaction and therefore requires the pay station to be connected to a network. The advantage of all these solutions is the elimination of cash and, in particular, small change.

Smart cards offer a further option: to use a 'personal parking meter' (see Figure 16.4) incorporating a real-time clock and display, placed within sight inside the vehicle. Since users must usually buy their own device, this is most suitable for areas where parking is both restricted and paid for.

16.3.5   Air travel

Airlines have been experimenting with electronic ticketing for many years. The 'e-ticket' (where all information on the booking is stored in the airline's central system and accessed through a reference, such as the credit card used for the booking, a frequent flyer card or a printed bar code) is now widely used and many airlines are able to share information on e-tickets.

Tickets and reservation data may also be stored in a smart card: this is defined by the International Air Transport Association (IATA) Resolution 791. An e-ticket or reservation could, therefore, be downloaded to the card over the internet or to a mobile phone (a boarding pass must still be issued to meet Warsaw Convention rules). In the 1990s several airlines experimented with contactless cards for use in ticketless travel; for example the German airline Lufthansa issued 130 000 of its 'Miles and More' frequent flier cards with this function. However, in practice airlines have found much greater advantages from centrally stored and interoperable e-tickets, often in combination with a two-dimensional bar code. Lufthansa has dropped the ticketless travel function and reverted to magnetic stripe for its 'Miles and More' cards, including those with credit-card functions.

However, if the card includes a biometric or other card-holder authentication then this considerably enhances the proposition: airlines are continuously balancing the needs of security against the convenience of the customer, and the combined ticket-identification card represents the best of both worlds. For some time in the early 2000s, Visa and IATA worked towards a co-operation on standards, but no agreement was reached. We will see in Chapter 17 that airports and private companies are now moving into the space left by the airlines.

16.4   Card and system requirements

16.4.1   Standards

As already mentioned, most smart cards used in transportation are contactless, and will meet either the ISO 14443 standards (type A or B) or a similar proprietary standard such as Sony FeliCa™ for the communication between card and reader. There is growing interest in NFC (ISO 18092) in this sector, and many readers are designed to read NFC as well as ISO 14443 cards.

However, this relatively tidy position at the lower layers is not matched when we look at the data, messaging and security structures used in the different transportation sectors. For public transport, the largest schemes still use one of two proprietary card types: Philips MiFare™ or Sony FeliCa™. The data structures and applications used with these cards are not standardised, and because they both use symmetric encryption it is not generally possible for cards from one scheme to be used in another.

There are several initiatives within the sector that address the need for interoperability, although not all operators feel the need to work with other cities or schemes. In Europe, *ITSO* started life as a consortium of UK local authority transport authorities, but has now evolved into a much more general smart-card interoperability scheme, with options for a very wide range of card types. ITSO terminals make use of a SAM to store top-level keys for all the relevant service providers, which enables a card issued by one operator to be used in another area by loading the relevant ticket product onto the card.

Calypso is another interoperability standard, this time originating in an EU initiative and, again, covering more than public-transport ticketing. Calypso specifies each of the layers of a contactless card/terminal interface, from the low-level communication standards, through the data structures (which correspond to an older standard,

EN 1545) to the terminal application and security architecture. Calypso cards have several functions that facilitate multi-operator acceptance, including a strong security and encryption scheme, but the standard does not include the back office systems that are included in ITSO.

ITSO is mostly used in the UK, while Calypso is used in several European cities, including Paris, Lisbon and Venice; they can each recognise each other's cards, but with a limited range of interoperability. The two groups of standards are working together to define a wider range of common functions.

Road tolling falls within the wide field of road traffic telematics. Digital short-range communication is standardised in EN 12253:2004, which along with many of the related technologies and applications is managed by TC278: Road Transport and Telematics, of the European Standardisation Centre Information Systems group CEN/ISSS [3].

Chip-card storage of airline e-tickets is specified in IATA Resolution 791; again there are several related resolutions that govern e-commerce in the airline industry, all of which are managed by IATA. Additionally, all cross-border transportation is affected by international security and identification requirements; these will be discussed in more detail in the following chapter.

16.4.2   Security

The security of the card itself is a growing issue: many cards used in public-transport schemes were designed for use in a closed system and where the marginal cost of an extra passenger is not significant. Where transaction time is important, most operators have traditionally chosen security schemes that allow rapid, offline transactions – typically a symmetric key scheme with a SAM in the gate or reader. However, with the growing interest in added-value services and in interoperability, the security of the card and system becomes much more important: value matters more when it is owed to a third party, while for e-money systems financial services regulators will insist on 'best practice' security to protect the value.

Under these conditions, single-length DES keys (56 bits) or proprietary algorithms are not usually regarded as adequate protection for monetary value, while some systems have known weaknesses (e.g., a point in the messaging cycle or an interface where keys or values are available in cleartext form). Most microprocessor (rather than wired-logic) cards can meet the higher level of security likely to be required for multi-operator use or for storage of open monetary value, although the system security strength is not determined by the card alone: other parts of the system must also be correctly designed. For the best combination of security and transaction speed, however, a crypto-processor on the card may well be required.

16.4.3   Dual-interface cards

As discussed in Chapter 3, there are advantages and disadvantages in using contactless interfaces. Where a card will be used for transport ticketing only, there is probably little need for a contact interface, although if the card may be updated by the operator (e.g., renewing an annual season ticket or changing personal details on the card), a contact interface may be preferred by the security department.

When we start to add other applications on the card: for example, verification of card-holder ID, an electronic purse or a driving licence, it is probable that the primary use of the card will remain contactless (otherwise there is not a good fit between the applications). However, in many cases the initial loading and any subsequent updating of the data for the additional application should be through a contact interface, for reasons of both security and reliability. For this reason, many transport cards do support both contact and contactless interfaces, with some functions permitted only through the contact interface.

16.4.4 Operational aspects

Many card-based public-transport schemes require the system to be fully gated, with turnstiles at exits as well as when boarding; this is expensive in capital cost and can be difficult to design (for example, if buses form part of the network it is difficult to enforce the use of the card when leaving the bus). Fare structures must be designed to handle the limitations of the check-in–check-out structure. It must be possible to open the gates for safety reasons (for example, in case of fire or panic) and the card or system must recover from such incomplete journeys.

In many systems incomplete journey records may arise for other reasons too: data collected at gates and on buses are only uploaded to the host system at intervals (often only at the end of the shift or journey); this process may fail or the data may be corrupted. The system may be able to reconstruct some journeys from other information, but in other cases it must simply delete the journey record or assume the minimum possible journey.

Most systems enforce some rules, for example a 'no passback' rule that prevents the same ticket being used by several people, or time bands in which certain tickets cannot be used. Other rules (for example, forbidding the use of someone else's ticket, or an adult using a child ticket) are difficult to enforce automatically in a high-volume environment and must be backed up by an operational control.

Once a public-transport card has been issued, it is difficult to recall it or update it, other than when the card is being used. Many systems include an 'action list' facility, in which the host system sends to all terminals a list of cards requiring an update. Small changes can be carried out during a boarding or gate entry transaction by using a script, but for more significant changes, the user may simply receive a message asking them to go to a ticket office to have their card updated.

16.4.5 Upgrading systems

The first public-transport smart-card schemes were installed in the mid 1990s, using the best technology then available. In many cases these systems continue to perform extremely well, however there does come a point at which the capacity, security or functionality of these systems must be upgraded.

There are several card products available that emulate the legacy wired-logic cards, as well as permitting new applications. In most cases they offer full ISO 7816 and 14443 support, but with enhanced security, including crypto-processors. An operator can issue new cards alongside the existing card estate, and these new cards may be read by existing readers and terminals. The terminals can then be upgraded gradually to include the new functions; when this process is complete, the operator has the option of issuing cards without the emulation function.

16.5   The future of multi-application cards in transportation

For many years, ticketing and payment were afterthoughts in much transportation system design. This has changed dramatically and revenue protection is now a key function in most transport providers' organisations, with cards playing an important rôle.

Whereas large-scale public-transport systems in major cities have been leading the way in their use of smart cards, smaller operators and other modes of transport are now embracing the technology at some speed; as generalised systems become available, the barriers to entry are falling. Multi-modal transport cards are likely to be one of the first application groups to be widely shared across organisations and organisation types, and it is possible that the experience gained in this way will help transport companies and their card system operators to be the lead issuers for many contactless card groupings.

16.6   References

[1] *Electronic Toll Collection*. Intelligent Transportation Systems, California Center for Innovative Transportation at the University of California at Berkeley. www.calccit.org
[2] Pasquali, F. *Relationships With the Motorway Toll Collection Schemes: the Cases of Florence and Rome in the Experience of Autostrade S.p.A.* www.transport-pricing.net/download/rome.pdf
[3] www.nen.nl/cen278/

17 Government and citizens' cards

Any card issued by a central or local government is liable to be branded as an 'identity card'. In many Western liberal countries that poses automatic grounds for suspicion of the issuer's motives; this chapter explores some of these motives and the issues surrounding government-issued cards.

17.1　Databases and cards

All card systems depend on a central database in some form. But for government cards in particular, it is important to distinguish card projects from the databases that underlie them. The growth in government ID card projects has been accompanied by growing concerns, from civil liberties groups in particular, about the potential loss of privacy these projects entail, and the potential for abuse and discrimination.

In practice, the use of large-scale *databases* is expanding strongly and does offer some scope for abuse; a correctly implemented card system linked to these databases offers the potential to control access to the data and give individuals more power over the way their own data are used. It is ironic that much of the opposition to identity cards implies that the use of a *card* represents an infringement of privacy, whereas a well-implemented card system should actually help to manage the privacy risk and to give citizens a degree of control over access to their records that they are unlikely to gain without such a card.

Often, in the examples below, the benefits are more attributable to the database than to the card, but the card significantly adds to the value of the database, by making it secure and publicly acceptable.

17.2　Electronic passports

The high volume of people crossing borders places a premium on efficiency in performing passport checks. Machine-readable passports, using optical character recognition, have reduced the time required for capturing personal details on entry to, or exit from, a country, while the use of a common format for these data means that airlines are able to send passport details to the destination country in advance.

Further time savings could be achieved if other parts of the process could be automated. Once their identity can be established, many passengers could be allowed through with minimal formality, leaving only those requiring a visa, for example, to be checked manually. This type of 'fast lane' operation has been successfully trialled with regular users at several airports.

In the USA, several private organisations are registering individuals and issuing them with cards that allow them to use special lanes at some airports. Currently such a system operates at Orlando International Airport in Florida; the company involved [1] is seeking approval to install lanes at Chicago O'Hare and New York JFK Airports.

As well as saving time, card-based systems offer greater accuracy: in Chapter 5, I showed that manual checking of identity – using photographs, signatures or other visual means – can rarely, if ever, achieve a sufficient degree of accuracy to detect impostors without also challenging many valid users. In practice, passport checks at borders are more effective than this would suggest, because those inspecting the passports are able to spot changed or forged documents and behaviour patterns associated with masquerade and dissimulation, but many forged or altered documents do, in fact, pass through.

Both the time-saving and accuracy aspects are addressed by the programme being undertaken under the auspices of the International Civil Aviation Organization (ICAO) and involving passport authorities, airlines, passport and smart-card manufacturers as well as biometrics companies. The ICAO is also working on other types of machine-readable travel document (MRTD), including an e-visa (issued to citizens of another country).

Interoperability is of the essence for an e-passport: each passport must be readable at all borders. Other key requirements include:

- *Technical reliability and durability*: documents and systems must last the full life of the travel document, often ten years. Cards must not fail when the holder is at a border crossing thousands of miles from home, even though they may have been subjected to some rough handling.
- *Practicality*: solutions must be operationally feasible in a wide range of environments and not require a range of different systems and equipment.

The ICAO has followed an iterative sequence of interoperability tests, pilots and specifications, resulting in a standard for machine-readable passports using biometrics [2]. This uses contactless smart-card technology (ISO 14443) but allows alternative formats, including a card or a chip and antenna embedded in the cover or an inside page of a passport. For reasons of sovereignty and authentication control, e-visas will in most cases not be stored in e-passports, but on separate cards.

The testing process has now resulted in acceptable levels of interoperability (although most suppliers still have some problem to resolve) [3] and many governments are now starting to issue chip-based passports.

17.3   Identity cards

Cards also offer the potential for improved efficiency in delivering government services, which have long been a byword for unnecessary form filling and duplicated effort. In managing government budgets, too, there is often scope for reducing over-spending and fraud, for example by accurately identifying benefit claimants.

National security is often quoted as part of the business case for issuing identity cards,[1] however, a detailed risk analysis shows that in many situations the security benefits would be marginal, while there is a risk that using the card might reduce the incentive to perform other, possibly more effective, checks. The existence of a com-pulsory ID card scheme does, however, create one additional barrier in a multi-layer security scheme, and may permit a controlled integration of databases that does have security benefits.

Countries that have implemented smart-card-based national identity cards have generally regarded them as very successful. They have usually replaced simpler printed, magnetic-stripe or bar-coded cards; the smart card not only reduces the scope for counterfeit cards but also allows further functions such as a driving licence or a digital signature.

However, in several other countries the debate continues: it is particularly heated in Australia, where an earlier (1987) proposal for an 'Australia card' was abandoned in the teeth of fierce opposition. The current government is convinced that the advantages of its proposed Access card for benefit payments and public administra-tion would far outweigh the drawbacks, and is conducting a campaign through the press to garner public support for this view, while opponents of the card have advanced a wide range of counter-arguments, most particularly around the cost (A\$1.1 bn) and manageability of the project.

Cost has also been the focus of the argument in the United Kingdom, where the Identity Cards Act 2006 provides for an ID card that will be issued alongside passports. Although initially voluntary, it would become compulsory by 2013. The card itself will contain three biometrics: two or more fingerprints, a digitised facial image and an iris scan. It is linked to the National Identity Register, which, in addition to the biometric data, will hold the present and all registered past addresses of card-holders. Although the database will be drawn together from existing databases, the creation of a national identity database remains highly contentious [4].

By contrast, the need for national ID cards is accepted without question in more than half the countries in the world, and many are now migrating those cards to a smart-card platform. These include Belgium, Denmark, France, Italy, Switzerland, Saudi Arabia, UAE, China, Thailand, Malaysia, Singapore and Hong Kong while others, such as the Philippines, are running pilots prior to full rollout. Two of these projects are described in the case studies in this chapter.

---

[1] See, e.g., UK Prime Minister Tony Blair in Parliament, 17 May 2005.

Such cards are usually issued by a dedicated government agency, which controls the cards and also permits other government departments to use and manage their own spaces. The existence of a clear hierarchy and, in many cases, a specific law governing the issue and use of the card, is a major factor in the success of multi-application card projects in this field.

Case study K – The Sultanate of Oman – national ID programme

by Martin Arndt, Project Manager, Royal Oman Police

Organisation

In October 1999, His Majesty, the Sultan of Oman passed royal decree 66/99; in this legislation he laid out his plan for enhancing the national ID programme. This included the creation of the Directorate General of Civil Status (DGCS) within the Royal Oman Police (ROP), as the owner of the project. The DGCS went through a tender process and in October 2002 selected Gemplus to implement the first National Registration System and the first smart-card-based citizen ID programme in the Middle East.

Business requirements

In line with the modernisation strategy for Oman pursued by the Sultan and the Omani government, this project had a few main goals:
- Modernise the National Registry System (NRS);
- Simplify and speed up the administrative processes;
- Provide better qualitative public services to Omani citizens and residents;
- Engage and promote the usage of IT technology and e-government facilities;
- Increase Homeland Security.

Solution

Technology
A Gemplus multi-application PKI-ready JavaCard was selected as a versatile platform to ensure that future applications could easily be implemented. The system includes biometric fingerprint minutiae technology provided by Sagem to allow authentication by portable terminals and airport immigration checkpoints.

The Royal Oman Police also decided to streamline the ID-card production, to make the process for the Omani Citizen as fast and efficient as possible. This means that the personalisation and issuance is carried out in 12 geographically distributed centres. The personalisation equipment is provided by DataCard and the overall IT system by GBM (IBM local business partner) and the ROP.

Operations
As part of the project process, Gemplus had to carry out an extensive skills transfer and provide training to the ROP IT department, to ensure a successful handover as well as maintaining high levels of system availability and resilience. It was always

the intention that the ROP would take over the system and Gemplus would remain involved in the maintenance.

Implementation
The first card was produced in November 2004 for the Sultan, and card production for Omani Citizens and residents commenced in January 2005. This was the culmination of two years of planning and implementation by the Gemplus and ROP project team. The rollout of the distributed personalisation and issuing centres was phased over 12 months, with one more centre opening by the end of 2006.

Challenges encountered
The DGCS decided to launch the card with three applications:
Identity – personal data including biometrics;
Border – border control application;
DL – Driving licence application.
It was the integration of the business process of the driving licence system and national ID card that proved challenging, with issues such as expiry dates and driving licence confiscation. The main issue was caused because the driving licence was printed on the ID card for interoperability with other Gulf countries. The problem was solved by changing business processes, and making the DL application optional.

The distributed personalisation and issuing centres required more resource and time commitment than if a centralised system had been used; extra resources had to be allocated during the project.

Outcome
Currently the system is producing 40–50 thousand cards per month, and 800 000 have been produced up to the end of June 2006. The Sultanate of Oman has achieved the goals that were laid out in the royal decree, and there is still much more potential to be realised in this multi-application ID card. There are plans to add other applications onto the card after it has been issued, as well as utilising the strong smart-card authentication at borders and government agencies.

17.4   'Cartes ville'

Municipalities and local authorities may issue cards for one of two reasons:
- To connect with their own citizens, improving administration and giving access to local facilities. Such cards are typically issued free: to the elderly, to students or to all those registered for a particular service.
- In tourist areas, to encourage visitors to use a wider range of local services. Such cards are more likely to be sold to visitors, or a cheaper printed card is used simply to give discounts.

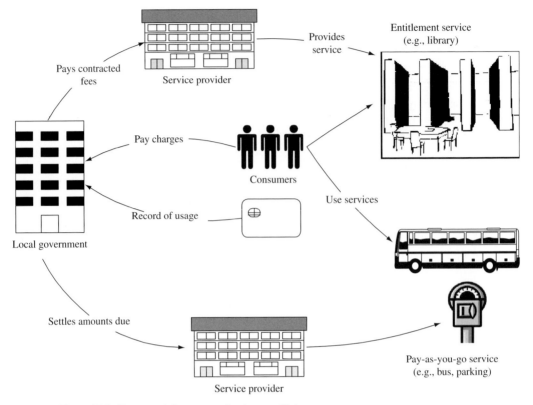

Figure 17.1 Commercial structure for 'carte ville'

Both types of card are often known by their French name meaning 'town cards', but the larger and functionally richer schemes fall in the latter category. A *carte ville* usually identifies the holder, status (e.g., student or retired) and a link to a record or records on the municipal database. It can, therefore, allow direct access to municipal facilities (such as libraries and sports facilities) and services (child care, school meals, home help, etc.) for those entitled to free services.

Most also have some form of stored value: for parking, entry to sports facilities or events, etc. The municipality runs the entire card scheme, although it may make payments to third parties for the services provided. This represents a very simple and effective commercial model for running a multi-application card (see Figure 17.1); however, the municipality also carries the entire financial cost and risk, and so such schemes are rarely very large, nor do they offer a wide range of services.

17.5   Health cards

Smart health cards can perform a number of functions:
- Provide proof of entitlement to treatment, either through membership of a national or employer scheme or by having paid premiums to a commercial insurer: different

levels of membership or premium may entitle the user to different levels of treatment, private facilities, etc.;

- Store emergency health information, such as blood group and allergies;
- Facilitate controlled access to medical records, which may be stored centrally or, more often, distributed across general practices, clinics and hospitals: in these schemes, including for example France's 'Carte Vitale', access to records requires both the patient's card and a health professional's card – the level of access granted depends on the qualifications and specialisation of the professional;
- Store a record of treatments, either for commercial or administrative reasons or to assist in further treatment;
- Monitor treatments, particularly for chronic diseases, and form part of the chain for transmitting these data to the doctor or clinic responsible: such cards are currently most often separate from other health cards.

Each of these functions normally requires a separate data file, each with its own access controls. While proof-of-entitlement cards are generally accepted without question, many of the other functions of health cards can be surprisingly controversial; although they are designed to improve treatment and benefit card-holders, and all the data are generally held in databases today, privacy lobbies are often concerned about the security of data held on the cards. There are particular sensitivities where the card is owned by a private health company, such as a health maintenance organisation (HMO) or by a private hospital or clinic, since it is often felt that the data may be used to justify higher pricing for people with specific diseases.

It is, therefore, essential to separate the administrative and commercial uses of the card from any disease-related or treatment-related functions, with only medically-qualified personnel accessing the latter on a need-to-know basis. People with serious, painful and chronic conditions are usually less sensitive to these issues than those in good health, and so more cards, or cards with more functions, are often issued to these groups (who are also heavy users of health services).

It seems to be extremely difficult to combine health cards with other smart-card applications; a very limited function (e.g., a health record number, and possibly some emergency health data) may be combined with other government-card functions, but further integration is limited: users mentally separate their health and health needs from other administrative and commercial needs, and health-card issuers must be able to demonstrate that no unqualified person can access any medical data.

Table 17.1 compares some examples of existing health card schemes and shows the wide range of options that exist.

The European Health Insurance Card, introduced in 2004, is a proof of national health insurance valid in all countries of the EU. The initial version is a printed card only, with a reference number that links to a database in the holder's home country; however, the intention is to replace this with a chip card containing, as a minimum, a patient identifier and emergency health records. These cards are due to be introduced in 2008; in the meantime, work is proceeding on adopting common standards for patient identifiers and a health record architecture that does not force member

Table 17.1. *Comparison of health card schemes in Germany, France, Malaysia and Taiwan*

Country	Card	Issued by	Functions	Card type	First issued	Planned evolution (2006 onwards)
Germany	Versichertenkarte	Insurance companies	Proof of insurance.	Protected memory	1993	Microprocessor card; treatment and partial medical history on card.
France	Carte Vitale	GIE 'Sésam-Vitale': service company set up by national and commercial insurers	Facilitate automated flow of paperwork between doctors, patients and insurance companies; Administrative data on holder and insurer.	ISO 7816 (native OS) with 8 kB memory	1998	Photograph on card; emergency medical data on card; larger memory.
Malaysia	Function on MyKad (national ID card)	National Registration Department	Next-of-kin contact details; Detailed information on lifetime health history, hospital admissions and (at doctor's discretion) recent treatments; Can only be read by medical professional.	Multi-application card, 64 kB memory	2001	Other MyKad applications.
Taiwan	National Health Insurance Card	Bureau of National Health Insurance (BNHI)	Personal data, including photo; Proof of entitlement (status, premiums paid, number of admissions); Emergency medical information (allergies and ongoing treatments); Immunisation records and organ donation; Cards can only be read in BNHI readers, which incorporate a SAM.	ISO 7816 (native OS)	2003	Further medical records; PIN protection for data.

countries to change their underlying health records system. A multi-national pilot (NETC@RDS [5]) is under way to test interoperability of various combinations of data sharing using both smart cards and host-based systems.

17.6 Student cards

Student cards, issued by schools and local authorities, also have a range of purposes. In the next chapter, I will address campus-card functions such as identification and access control, but from a government point of view student cards can assist in:

- Monitoring and recording attendance at classes: having students and pupils record their attendance eliminates the need for a manual 'roll call', which saves teacher time and improves accuracy. An automatic system can spot discrepancies and irregularities, and trigger appropriate action, while the card can accumulate attendance records from different institutions and activities, which may not all be linked to the same school management system.
- Paying for school meals: while many studies, e.g., [6],[7], have shown that regular, healthy meals improve pupils' behaviour and ability to learn, schools are often stuck in a conflict between cost management and freedom of choice. Using a smart card to pay for school meals offers many advantages, for example:
  - o It ensures that money given by parents for meals is not spent on other things;
  - o There is no scope for discrimination between those who pay for their own meals and those who receive government assistance;
  - o Pupils can be rewarded for healthy eating;
  - o The cost of collecting money is reduced and cost control is improved.
  A study carried out for the UK Department for Education and Skills and the Food Standards Agency [8] recommended that smart cards should be used to help incentivise healthy eating and to allow monitoring.
- Online access: in order not only to reduce costs but also improve pupils' technical literacy and understanding of information retrieval, many governments and educational establishments make information available to students online, and encourage them to make choices through this channel. It is much easier to ensure that students gain access only to the appropriate and relevant material (whether on the public internet or on an Intranet) if a full profile of the student is stored on the card, together with a password or access code (but see also the Shibboleth case study in the next chapter).
- Incentives: some governments have introduced national or local schemes that use smart cards and loyalty techniques to encourage students to attend courses or carry out other desirable activities. For example, the Connexions card in the UK [9], which is issued to 16-to-19-year-olds, works in partnership with national and local businesses to give rewards in return for points gained through attendance or achievement on academic and training courses or for voluntary work. The objective is to increase the number of children continuing in education after the minimum school-leaving age.

17.7   Additional functions on government cards

17.7.1   Proof of age

Young people often need to prove their age, often before they have a driving licence. A smart card, issued by a government-approved agency, offers a proof of age that is not easily forged or altered.

17.7.2   Driving licences

Driving licences must not be easily forged or altered, and it is also an advantage if they are machine readable. In the Indian state of Gujarat and in San Salvador, all driving licences are smart cards, while in Japan and some Chinese cities, a smart card is issued alongside the driving licence to record traffic violations and fines [10]. Many European countries are moving towards adding a chip to licences (for which there is already a standard card format); it has taken many years of discussion at EU level to converge the needs of all member states [11], but in March 2006 an agreement on driving licences left the chip option open.

Where a national ID card is being issued, however, as in Malaysia or Oman, there is a clear benefit in including a driving licence application on the card. Driving licences are often used informally as ID cards, and the combination of a licence with a cardholder verification method and government authentication renders the card more useful to both holder and checker.

17.7.3   Access to government services

Smart cards can facilitate efficient delivery of government services in three main ways: by permitting direct access to records and services through the internet or kiosks, by ensuring that officials can go directly to the correct record (and not, for example, to another person of the same name), and by reducing the need for repetitive form filling.

To be used in this way, the card must contain cross-references to all relevant identification numbers and case numbers, as well as basic information such as name and address, sex and age. Suitable controls are needed to ensure that these data are only available to the relevant applications. People moving house or changing their address represent a significant burden and opportunity for errors for central and local governments. Using a card that has references to many departments allows the citizen to update his or her own records and to ensure that the changes are propagated through the whole system. Again, controls are needed to ensure that this is not abused, for example, to divert mail to another address or to avoid legal enforcement.

Card-based services of this type are open to the criticism that they are discriminatory: they cannot be used by the illiterate and may often be difficult for those with sight problems, shaking hands or other handicaps. Cards should be used for facilitation but rarely as the only means of access to services.

Case study L – A UK local authority citizens' card scheme

by Richard Poynder, Chairman, Smartex Limited

Organisation

Many people regard local authorities (LAs) – local government bodies – as the sector with the greatest potential for the implementation of multi-application smart-card schemes in the UK. As most LAs are introducing similar functions within their citizens' card programmes, this case study incorporates use of the National SmartCard Project (NSCP) – the framework now being widely adopted.

Background

Since about 1997, LAs have recognised the potential benefits of smart cards issued to those who live or work within their catchment areas. Local authorities were attracted by the concept of providing people with a single card capable of performing a number of functions relating to the delivery of statutory council services. Many LAs also wanted to promote social inclusion and avoid stigmatisation of those who were entitled to free or subsidised benefits, such as school meals, access to leisure or concessionary travel.

In 1999, the UK government endorsed local-authority use of smart cards under the 'Modernising Government' programme. Central funding was provided to a small number of initiatives, leading to the development and publication of the National SmartCard Project (NSCP), which now includes a licence-based starter pack for LAs to use in establishing their card-holder databases and card fulfilment and personalisation services. Under the e-government programme, English local authorities are mandated to offer a wide range of their services electronically. Scotland, Wales and Northern Ireland have similar programmes, broadly following the English model.

Although many central government initiatives (such as national ID, driving licence and health cards) have little impact on LAs, one that does is the Department for Transport's support for the ITSO specification (see Chapter 16), since this application has a direct interest for LAs in their provision of concessionary transport facilities. Some local authorities have commenced their smart-card projects with transport functions – typically based on 4 kB MiFare™ contactless cards – with a general ambition to incorporate other citizens' functions within them, whilst others have deployed dual-interface citizens' cards first, and are now trying to incorporate the ITSO shell within them as a later phase.

Business requirements

Whilst not businesses in the commercial sense, most LAs share underlying goals for their smart-card programmes, including:
- Fulfilment of a policy of social inclusion;
- Avoiding stigmatisation of citizens entitled to statutory concessions;

- Reduction in administrative paperwork for council employees;
- Reduction in fraud by citizens and council staff;
- Securing e-transactions;
- More effective service provision for citizens in rural areas;
- Meeting government modernisation objectives;
- More effective application of public funds;
- Consolidation of all aspects of citizens' relationships with their LAs;
- Increased convenience to citizens by combining all services within one card.

Most local authorities have recognised the need for their citizens' cards to evolve over time, and for additional functions and applications to be added in later phases. They have typically started with applications for concessionary (subsidised) transport, schoolchildren, library and leisure outlets. Many cards include generic functions for identity and stored value, while proof-of-age and incentivisation programmes are also becoming popular themes. There is some overlap with central government initiatives in these last areas.

Solution
The NSCP specifications are now offering a degree of standardisation and potential interoperability across councils, and local authorities in general are adopting similar approaches to technology, implementation and operations.

Technology
Most early schemes adopted a dual-interface JCOP card (originally preferred by the NSCP). This is still generally the technology planned within English local authorities, but Scottish and Welsh councils, choosing to concentrate first on concessionary public-transport applications, are initially deploying MiFare™ contactless-only cards. It remains to be seen whether these wired-logic cards offer sufficient security or capacity to support the other applications required, or whether there will be a general migration either to microprocessor-based contactless cards or to dual-interface technology.

Most card readers for points of service within citizens' card schemes have, to date, largely been proprietary, with readers integrated within school and leisure outlet EPoS terminals. Libraries have frequently required bar codes to be printed on the card, although some have started to use the card's chip for this purpose.

The NSCP starter pack includes card-holder enrolment and card personalisation software, and LAs are increasingly using this for their card fulfilment requirements.

Implementation
Most local authorities expect a planning and implementation period of about 18 months. Few current implementations, apart from those commencing with public-transport functions, have as yet moved from the pilot phase to full roll-out; they have instead concentrated on the relatively easy captive markets within

schools, libraries and leisure centres. Such pilots have, therefore, not yet properly tested the overall utility of a voluntary, multi-application card deployment, and its effect upon council staff and the citizens whom they serve.

Organisation and operation

Whilst local authorities in general have recognised that the introduction of multi-application citizens' cards will affect operations within all participating departments, none as yet has undertaken the organisational and operational changes that should yield the benefits claimed for this technology. In some cases, LAs have set up separate operating companies to manage, control and own their schemes, but few are adding revenue streams by introducing third-party applications within their cards; commercial public-transport functions are the most probable initial candidates.

Several regional groups of LAs have recognised the benefits of working together, both in terms of interoperability and of economies of scale through bulk and standardised purchasing, but few have yet achieved these objectives. Scotland and Wales, with their more monolithic and central approaches to the introduction of citizens' card schemes, may be better placed than England to realise such benefits.

Problems encountered

Most local authorities that have piloted citizens' card schemes have, through these pilots, recognised the complex nature and far-reaching impact of the process. As well as introducing comparatively novel technology, these schemes have a major effect on a LA's administrative procedures and on the working practices of its staff. Inter-departmental co-operation is essential when a common database, card-holder interface and a single, multi-application card are introduced.

A wide variety of problems and challenges have emerged, including:

- Departmental silo mentality: departments used to controlling their own budgets, and often responsible for their own single-function card issue and associated databases, can find it difficult to give up some of that independence;
- Organisational inertia: staff and unions often react badly to any change in working practices;
- Lack of a sustainable business case: LAs often fail to find financial benefits to offset the cost of a citizens' card scheme and are unwilling actively to seek commercial application owners to share card real estate;
- Inability to define financial scheme boundaries: the introduction of multi-application smart cards often requires upgrades to underlying application systems, for example in schools and leisure outlets. There can be confusion and conflict as to whose budget carries the cost of these upgrades, while LA accounting systems often fail to allocate cost savings across departments;
- Complex tendering procedures: these can lead either to a lack of interoperability between the various components that make up a scheme, or to the appointment

of a single supplier that then attempts to lock the LA into a set of sometimes inappropriate proprietary technology;

- Lack of marketing experience: LAs find it challenging to deal with the complex branding and ownership issues raised by multi-application smart cards;
- Perceived legal barriers: data protection and e-money legislation have direct impacts on citizens' card schemes, and fear of these constraints and associated penalties can deter LAs from pursuing multi-application schemes, or from incorporating e-purse facilities within them;
- Unrepresentative pilots: most LA citizens' card scheme pilots are restricted to functions in schools, libraries and leisure outlets. Few citizens make use of all of these, the audience is often captive (so the voluntary nature of any scheme is untested) and they do nothing to enable the remote delivery of services by electronic means;
- Difficulty in providing remote services: one of the main objectives of citizens' card schemes is to enable citizens who live or work in remote areas to benefit from council services through the internet. Unfortunately, many of the target groups do not possess the necessary home technology, and so there is no advance in the provision of e-services;
- Confusion over multiple government initiatives: some LAs are put off by the growing number of unlinked central government smart-card initiatives for this technology.

Outcome

Although there are many pilot schemes in operation, local authorities in England are currently reluctant to initiate new citizens' card programmes, for financial, political and cultural – rather than technical – reasons. Single-application concessionary transport cards are now widely issued in Scotland, but on a more fragmented basis in England.

The number of true multi-application citizens' cards in circulation in the UK at the end of 2006 was less than 250 000 cards. Within five years, however, unless there is a change in government policy and reduced pressure in support of its modernisation programme, this figure is expected to grow to over 40 million cards across the 468 UK authorities.

17.7.4 E-commerce authentication

Governments have an interest in promoting e-commerce, but must also ensure that their citizens understand the risks of dealing in cyberspace and are able to protect themselves. For this reason, many governments have enacted electronic signature laws and provide, or promote the use of, cards that allow the general public to sign electronically, with the force of law. Digital signatures normally use public-key cryptography and so cards used for this purpose must normally have crypto-processors.

17.7.5   Payment

Although governments normally try to avoid direct responsibility for payments (this function is reserved to banks), they are involved whenever the payment is for a government service (which in many countries may include utilities or rent), or when they seek to control the way government benefits (usually not unemployment benefits or pensions, but more targeted benefits such as food and clothing allowances or milk tokens) are used. There are always problems in seeking to restrict the way money is used, but a card can at least provide a much higher degree of traceability and, thereby, reduce the level of fraud and 'leakage'. Governments may also promote functions such as e-money, to prime the pump and motivate commercial suppliers.

In most of these cases, the payment application itself may be identical to those described in Chapter 15, but instead of using a conventional network to 'clear' these transactions, the government performs that function itself. Since the government does not belong to any of the international banking networks, it must work through a bank (typically the national bank) to gain access to the authentication structures and intellectual property that these networks own.

Case study M – Malaysian Government multipurpose smart card (MyKad)

by Wan Mohd Ariffin, Government Program Director for MyKad 1998–2004

Background
In 1997, Malaysia embarked on a flagship project to deploy a chip-based multi-application national identity card to replace the laminated-paper-based identity card issued by the National Registration Department.

"Malaysia will have the first multi-application smart card in the world. Only one card, which contains identity card, digital signature and access to government departments, banks, credit facilities, telecommunications, transport and health services." (Dr Mahathir Mohamad, former Malaysian Prime Minister, in 1996)

Applications
The national ID application is the anchor application, as it is mandatory for all Malaysian citizens and permanent residents above 12 years old to possess an identity card. The card contains two biometrics: the face and a pair of thumbprint templates. These are captured during the card application process, and subsequently processed for storage in the chip.

The card was also designed to carry the Malaysian driving licence to replace the former laminated-paper document. The third application is the passport information file, which allows Malaysian travellers departing from and returning to Malaysia to use automatic gates (Autogates) at immigration checkpoints in Malaysia, avoiding queues at manned counters. The fourth government-based application is the health card. This application aims to provide a portable record of basic health data, both static and dynamic, as well as access to free or subsidised health care for the card-holder.

A novel feature of this card was the inclusion of a Proton[2] e-purse, branded locally as MEPSCash. This was, in fact, one of the earliest examples of a multi-application smart card that accommodated both government and banking applications, despite the very disparate business requirements of these two sectors.

After the card was launched, an automatic teller machine (ATM) application was incorporated. This was identified as a desirable feature to enhance the utility of the card and required the project team to meet the disparate requirements of the government and the banks. Finally, the seventh function incorporated into the card was the public-key infrastructure (PKI) component. This enabled the card to store the private key of an asymmetric key pair as well as the related digital certificates and allowed the creation of digital signatures in co-operation with Malaysia's two existing certification authorities (CAs).

Solution

A Malaysian-based company provided the card and the proprietary multi-application operating system secured in flash memory. The system enables the separation of the several applications in the chip and supports the PKI functionality with a built-in crypto-processor.

The card and all its supporting systems were developed over a two-year period; a fully functional version of the card was deployed in a field test in August 2000 and it was formally launched in September 2001. The card was officially named 'MyKad' in the course of the launch. The 32 kbyte chip used in the original MyKad was upgraded to a 64 kbyte chip in 2003.

In January 2003, 100 000 cards integrated with MiFare[TM] contactless modules were issued as a trial to allow the card to be used for mass transit transport and road toll payments.

Challenges

Initial risks

The Malaysian government accepted the risk of selecting a prototype design during the tender evaluation exercise in 1998 as it wanted the most advanced chip to be deployed when MyKad was issued in 2001.

Another major risk was the commitment to include the Proton e-purse into the card, although when the project specification was finalised in 1998, only one card operating system was Proton certified. Certification of a new chip and operating system was estimated to take two years, but actually took less than a year from the time the chip became available in engineering prototype form in early 2000.

---

[2] See Chapter 15.

Infrastructure and reading devices

The introduction of smart cards requires a complex and completely new infrastructure. The system may not be online but devices are required at all points of interaction. Several devices have been developed to support the deployment of MyKad.

A personal key-ring reader allows ordinary citizens to access the 'open data' (address, name, ID number, e-cash balance, etc.) stored in the card. It also allows reading of the driving licence details: type of licence, category and expiry date (this information is not available on the card surface). A larger and more powerful key-ring device allows enforcement officers to access the full repertoire of information stored in the card.

A mobile card-acceptance device equipped with SAM slots provides write capabilities to 'restricted areas' in the chip as well as thumbprint biometrics comparison and photo display. A PC-based device called the chip data verifier (CDV) allows the entire chip-stored data suite to be displayed and checked for integrity and accuracy (text details, colour photo, thumbprints, etc.). Government kiosks incorporating CDV functionalities allow citizens not only to check chip data but also to load passport information into the card (if not already done at issuance). Banking-style terminals enable financial institutions and government offices to capture basic information required in a paperless environment as well as to load or decrement the e-purse and access the ATM application.

Personalisation

Government procedures for the production of the national ID card dictated that the entire security personalisation of the card be done on government premises. This posed a considerable challenge to the project, as a new personalisation system had to be devised that also met the disparate requirements of the e-cash and PKI personalisation. Today, the National Registration Department of the Government of Malaysia has one of the most advanced and diverse smart-card personalisation centres in the world, capable of personalising 12 000 MyKads a day.

Inter-department co-ordination

One of the major challenges of the MyKad project was the difficulty for the several government departments involved in the project (National Registration, Transport and Immigration) to subordinate their respective data (and by inference their authority) to another department. The project team met this challenge by developing rules, procedures and technological tools to ensure that each department retained control over its own data. At the same time, procedures were also developed to share data that were common to and used by all departments.

Ensuring public acceptance

To encourage the public to change their old ID card into MyKad, the Malaysian government announced a free conversion offer valid up to the end of 2005. A comprehensive and widespread external communication programme was developed to introduce MyKad and its features and benefits to the public. The government also prioritised the identification of uses for MyKad and the deployment of these uses as soon as they were identified. All government-related transactions that required an identity card to prove identity were equipped with MyKad readers to automate and accelerate the processes. Special 'fast track' lanes were set up in government service counters to reduce queuing times and enable more efficient payment for services if MyKad was used.

Outcome

MyKad is now the legal identity card for Malaysians. By February 2006, a total of 18.3 million MyKads had been issued to the Malaysian public as their national identification cards. The system automatically inserts the driving licence and passport data into the individual's MyKad if the individual already has these documents.

In the USA, electronic benefit-transfer (EBT) cards are the main way for paying benefits of all kinds to those who do not have bank accounts. Although standard EBT cards use a magnetic stripe only, some states and programmes have added to their EBT cards a chip, either containing a biometric or for other purposes (e.g., in New Mexico a contactless chip is used to track social services' use of transportation services) [12].

17.7.6   Record of qualifications

Recent growth in both international mobility and the importance of qualifications has led to a rapid increase in fraudulent claims to qualifications, while specialisation in many jobs increases the importance of having the right qualifications. Several govern-ments have, therefore, started schemes for recording qualifications using a smart card. The records may be signed by the educational or training establishment, which in turn is approved by the government. These records may be used when candidates apply for jobs or promotions, when publishing or advertising services, or for Health and Safety inspections. An example of this is the building and construction industry in the United States, where the 'Smarter Skills System' smart card was introduced in 2003 [13].

17.8   Data protection and privacy issues

Although data protection and privacy are issues for all sectors and for nearly all smart-card schemes, they are particularly important for government schemes. Some countries have overcome this problem by creating a specific law to cover their ID cards but in others existing data protection laws are weak.

To resolve the problem, however, governments must not only ensure that data are genuinely only used for the purpose for which they were collected, and that data cannot 'leak' from one department or use to another; they must also be able to demonstrate or give confidence to citizens that this is the case.

In 2002 the Japanese Government launched the Basic Residents Register Network *JukiNet*, which linked national, regional and local government databases. Within hours, several local governments had disconnected themselves, citing inadequate privacy protection. However, a privacy law that had been pending for some time was rushed through the Diet and Juki Cards (chip cards containing name, address, civil status and a link to the holder's record in Juki Net) are now commonplace [14].

17.9 The future for multi-application government and citizens' cards

As the case studies in this chapter show, many countries have already developed multi-application cards that yield benefits for government administration. It is more questionable whether general-purpose central-government-issued cards have shown real benefits to citizens. But several local government and agency initiatives do yield benefits for citizens, including health cards and those aimed at schools or benefit-holders.

The challenge for central governments is to persuade citizens that the card is of benefit to them, for example by emphasising the protection of personal data and the desirability of controlling access to databases; this is likely to require applications demonstrably held on the card itself, rather than in any central system. In time, such schemes will probably become universal, although possibly in quite different formats in different countries, and requiring different combinations of applications and features.

17.10 References

[1] www.verifiedidentitypass.com/

[2] *ICAO Document 9303: Machine Readable Travel Documents, Biometrics Deployment of Machine Readable Travel Documents 2004*. Montreal, International Civil Aviation Authority 2004. http://mrtd.icao.int/

[3] *e-Passport Interoperabiity Test Event, Berlin 29 May–1 June 2006*. Deutsches Institut für Normung 2006. www.secunet.com/berlin/

[4] London School of Economics. *The Identity Project. An Assessment of the UK Identity Cards Bill & its Implications; Interim Report*. LSE March 2005

[5] www.netcards-project.com

[6] Nutrition: Eat up your veg. *Nature*, **221** (5179), 401, 1969

[7] Trichopoulou, A. *et al*. Dietary patterns and their socio-demographic determinants in 10 European countries: data from the DAFNE databank. *European Journal of Clinical Nutrition*, **60** (2), 181–90, 2006

[8] Nelson, M. *et al*. *School Meals in Secondary Schools in England*. King's College London: National Centre for Social Research. DFES Research Report RR557 2004

[9] www.connexionscard.com

[10] Woodward, K. Chip-based driver's licenses move forward, but in the slow lane. *Card Technology*, October 2005

[11] *Proposal for a Directive on Driving Licences.* European Commission COM (2003) 621/F. October 2003

[12] Espinosa, J. M. *et al. Developing the Client Referral, Ridership and Financial Tracking (CRRAFT) Transit Management System: CRRAFTing a Bridge to Coordinated Interagency Transportation.* 82nd Transportation Research Board Annual Meeting 2003.

[13] *Smarter Skills and Security System.* Building and Construction Trades Department 2006 (www.buildingtrades.org)

[14] *Diffusion of the Basic Resident Register Network (Juki Network) and Residential Smart Card (Juki Card).* Japan Productivity Center for Socio-Economic Development November 2004

18 Campus cards and closed user groups

For many of the card schemes discussed so far, the card has as much marketing as operational value for its issuer; it ties the consumer to the issuer and provides a channel for delivering services and differentiation. This chapter is concerned with a group of applications where the card-holder is already a member of a defined group, and the aim of the card is often to raise barriers around the group and prevent infiltration or abuse of the privileges of the group.

These include not only employee card schemes, schools and universities, but also holiday camps and clubs, prisons and detention centres. Athough they are often referred to as campus cards, the scope may be much wider than a physical campus or group of sites.

Sometimes principles that apply to public schemes must be completely rewritten for this environment: for example, employees may accept the storage of personal data or a biometric on the card as a condition of employment. Laws governing payments in legal tender may not apply to canteens or on-site vending machines.

For this reason it is often difficult to mix campus cards with open applications: for example, several banks have found it difficult to act as a card issuer for an electronic purse on a university card, since they are subject to regulation that imposes high standards and costs, and that makes it difficult to meet the rapid turnaround required to address lost cards in a university environment. From the university's point of view, the partnership limits the ability of the university to extend the functions of the card, while the effort required to meet the requirements of an external issuer may not be much less than carrying out the work in-house.

18.1 Identification

The core function of most campus cards is to act as an identifier and as proof that the holder belongs to the relevant community. As with a citizens' card, it provides a reference that allows access to database records: administrative, employment or academic. Some of the underlying data may well be sensitive, and so systems are designed to require controls on the person accessing the data as well as the subject. Many cards will carry a printed photograph, and security is improved if there is some form of password or machine-readable biometric in the card as well.

Older cards often relied on their coding scheme, or simply on the limited availability of readers, to provide security. This is no longer adequate for any realistic scheme; simple authentication and signature schemes are very easy to implement and are effective in preventing most alteration of data as well as unauthorised access. The issuer may act as its own certification authority, usually buying in an appropriate package, or may use a third-party service. Symmetric encryption is usually adequate for single, closed user-group schemes.

18.2   Physical access control

Contact-based smart cards are too cumbersome for most access control applications; however, they may be more appropriate in high-security environments, and particularly when the use of the card is combined with a biometric check. The card can remain in the reader while the check is carried out.

For most other access control, contactless smart cards are preferred, and are now starting to displace older Wiegand cards and RFID tags that only transmit a reference number to the reader.

One of the major advantages of the smart card in this environment is flexibility: the card may contain several card-holder verification methods for use at different times (e.g., card only while a door is attended, card + PIN at night, a biometric for access to a secure area, and a further password or fingerprint for computer system logon). It can also contain back-up evidence, in case the card-holder forgets his or her password, or request different combinations on an unpredictable basis to prevent replay attacks.

A smart card is preferable to a tag when any form of offline check is needed (for example, the car park barrier need not be connected to the online database but relies on data in the card only).

18.3   PC and systems access

It is now widely acknowledged that the old way of accessing computer systems, using a user identification (user ID) and password, is hopelessly insecure: people choose weak passwords, write them down or tell them to others, while even the strongest passwords can be captured, replayed or obtained from system files.

Any system with more than trivial security needs should use two-factor authentication: a combination of an intelligent token with a password. Smart cards are one of the easiest forms of token to implement and are available in several packages, either as part of an operating system or as a stand-alone security system. The system verifies the authenticity of the card and the card verifies that of the user. Time-variable or random functions increase the strength of the system by avoiding replay attacks.

The card can also contain a profile of the user and the many further passwords required to cross firewalls, access different websites or authorise payments. Careful spreading of these data between the card and central system should provide an optimum balance between security and convenience.

Case study N – Shibboleth

Organisation
Shibboleth is a project within the 'internet2' next-generation internet project, which is run as a consortium-based open source development within a number of US universities, partly funded by the US National Science Foundation. Shibboleth development has been co-ordinated by Ohio State University, which is also one of the first implementations.

Business requirement
Shibboleth addresses the need for computer users to access information on computer systems, other than their home networks, securely, without the need to register with each network. It focuses initially on the needs of students, researchers and staff in universities, particularly in state university systems, which share a wide range of resources.

It is designed to balance the needs of security (strong authentication of users where possible, and certainly for functions requiring personal authorisation and extended rights) with those of the user's privacy (there should be no need for the remote system to know all a user's details, only that they have a defined set of rights and permissions to use that system) and of user convenience (no need to remember a wide range of usernames and passwords). It forms a key part of the authentication and authorisation infrastructure (AAI) that underpins internet2.

Solution
Shibboleth is built on a federated or 'trust network' structure, in which organisations share data with other trusted organisations.

When a remote site asks for authentication (Where are you from?, or WAYF in Shibboleth-speak), the user's home network uses OASIS/SAML (Organisation for the Advancement of Structured Information Standards/Security Assertion Markup Language[1]) to send enough information about the user's credentials to allow them access to the target network's facilities. For example, these would usually include the person's affiliation (institution and department), status (student, researcher, professor) and (in a subsequent exchange) any additional rights negotiated for that person, such as the right to add records to a database.

Technology
Shibboleth does not, in principle, demand the use of a smart card: students, for example, may log on using a username and password. However, security

---

[1] www.oasis-open.org/specs/index.php

administrators everywhere are aware of the weaknesses of password-based systems (people write down passwords or use weak, easy-to-remember passwords; they share them with colleagues, forget to log off and are open to 'social engineering' attacks, where fraudsters obtain passwords simply by asking for them) and so many organisations insist on a minimum standard of security for their own systems, which typically includes two-factor authentication using a smart card.

The federated structure requires that an organisation seeking access to a resource for its users must meet the security needs of that resource (it must be able to make the assertions in its response). This encourages a virtuous circle in which all organisations must upgrade their security in order to have access to the most powerful resources (which on internet2 are very powerful indeed).

Issues encountered

The main issue for Shibboleth implementors is to ensure interoperability with existing single sign-on networks, so that users are not faced with a confusing array of sign-on methods.

Many existing smart-card authentication or sign-on tools make use of data stored on the user's smart card, typically a public key that allows the card to respond to a challenge with an appropriate cryptogram. However, for accessing remote sites they revert to passwords stored on the card in encrypted form. Shibboleth overcomes this problem but only for Shibboleth-enabled sites. Some combination may continue to be needed for some time.

Outcome

Shibboleth was first introduced in 2003, but its adoption has grown rapidly in 2005 and 2006. By the end of 2006, Shibboleth linked thousands of campuses in the USA, UK, Switzerland, Norway, France, Finland, Denmark and Australia; it has even demonstrated interoperability with the US Government 'E-Authentication Identity Federation'. As well as universities and research establishments, Shibboleth is being seen as a way for schools to access learning materials in a secure way and, hence, links with other school smart-card applications such as access control, attendance monitoring and school meals.

18.4   Authorisation and signing

A card can help to enforce a hierarchy or organisational structure, by giving card-holders rights and privileges appropriate to their rôle or position. It can also ensure that only those with the appropriate permissions are able to authorise changes, purchase orders, etc.

The use of a digital signature function supported by the card allows an audit trail of changes, orders and other reserved functions – although this sounds unexciting, a consistent audit trail of all such operations with full traceability to the card-holder is a very powerful tool for resolving disputes and ensuring responsible behaviour. Documents can also be signed electronically, which should lead to a reduction in paper movement and more consistent filing and retrieval.

18.5   Cashless payment

On many corporate, university or school sites it is possible to eliminate the use of cash completely from vending machines, canteens, bars, terminal and copier use, etc., through a stored-value application on the card.

Combinations of stored value and entitlement cards allow everyone access to services, either by right or by paying. As we noted in relation to school meals, it is not necessary for outsiders to know whether a given user is paying for a service or not.

The organisation reduces administration costs and cash losses; the equipment is less prone to breakdown or malicious damage, and accountability is greatly improved. For users, the card is usually more secure than cash or a bank payment card, and with a well designed system using the card should be quicker than cash. Whereas in the 1990s such systems were usually based on magnetic stripe cards or contact smart cards, the contactless cards and applications now available offer a much better balance of security and convenience.

Case study O – Gwernyfed High School

Organisation
Gwernyfed is a small (600 pupil) comprehensive school for boys and girls from 11 to 18, serving a wide rural catchment area in central Wales and over the border into England. The main school building is a historic mansion house, and a small proportion of pupils have Welsh as their first language.

Business requirement
Powys Council, which is responsible for the school, initially sought a bespoke registration system for Gwernyfed, while the headmaster was keen that it should be a unique solution that would push the school forward and make it the most technologically advanced school in the area.

Solution
Attendance monitoring
The school started by installing an off-the-shelf Smart Time and Attendance System from supplier Fortress GB. Each pupil is issued with a contactless smart card. The pupils use the cards to register the times of their attendances each day and for each class,

by holding their cards against readers that are attached to the wall of each classroom. Sixth-form pupils register their attendance through readers located in the corridors.

The raw data on each student (times of attendance on a daily basis) are sent, through a web-based interface created by Fortress for this project, to an external bureau that produces the necessary reports and sends them back to the school. The system also integrates with the Schools Information Management System (SIMS) used by most schools in the UK, which provides the data needed for initial personalisation of the cards.

Cashless catering

The catering department of Powys Council then decided to roll-out an e-purse facility on the cards to create a cashless cafeteria environment at Gwernyfed. Experience from other schools across the UK held out the prospect of shorter queues coupled with a substantial reduction in both theft amongst children and bullying of those children entitled to receive free school meals.

The Cashless Catering application is pre-loaded onto each contactless smart card, and so no recall or re-issuing of cards was needed. Readers were simply attached to the tills and a user interface through the tills' touch-screens was created. To pay for their meals, pupils hold their card against the reader attached to the till; the cashier then sees the pupil's name and year and whether the pupil is entitled to a free meal or not.

Money can be loaded onto the cards either through a parent's cheque handed in to staff at the school, or by payment via a secure internet website connection, over the phone or even directly to Powys Council. Many parents enjoy the peace of mind of knowing that their children are not carrying cash and that they can keep track of their children's spending at lunch, whereas older pupils benefit from the responsibility of loading their own RF smart card with lunch money.

Further developments

The Issuing Station, Time and Attendance System and Cashless Cafeteria are only three facets of Fortress' multi-application *Smart School Solution*. The solution is designed to allow schools the freedom to stagger the activation of applications as and when the school is ready, without having to re-issue cards.

Further applications include access control, registration (including SMS text messages alerting parents of their children's absences), printing and photocopying, interoperability with third parties (including local transport systems), stored value for use with vending machines, and increased parental involvement with the school through a secure website that details information on each child's attendance and e-purse purchases.

Experience has shown that cashless payment in schools and universities is successful where it can cover a high proportion of students' expenditure on site but is much less successful on city-centre campuses, where expenditure is more widely spread and students are likely to make more use of other facilities for travel and sport.

18.6   Operational requirements

For a typical closed user group, such as a university or corporation, it is usually possible to carry out all the card issuance, terminal management and transaction clearing in-house. Other organisations may find it preferable to subcontract some of these functions to a more specialised company.

18.6.1   Card issuance and card management

The starting point is a reliable database of all users. For most campus schemes, users are entered into the database outside the card scheme, by a personnel department, student management system or booking system. The database must contain sufficient information to populate the card parameters, with the exception of the cryptographic fields, which are computed by the card-issuing system.

Most card-issuing systems will nowadays incorporate a security module of some kind to store the organisation's master keys; this must be managed and maintained according to a *security policy*. It is important that there is one person or group responsible for setting and enforcing this policy, with the full support of senior management (who are sometimes the worst transgressors of security rules). The card-issuing system must, itself, be protected by suitable physical and logical access controls.

Where passwords will be used as part of the scheme, specific techniques must be used to ensure that users choose strong passwords and change them regularly. The use of long and random passwords is recommended only if the passwords will be stored on the card (in protected form) – otherwise users will simply write the password down. The use of password reminders greatly reduces the number of helpdesk calls.

18.6.2   Biometric enrolment

The successful capture of biometric data (fingerprints, photographs, etc.) is highly dependent on environmental factors, including the design of the booth, lighting, positioning of cameras (particularly if there may be disabled people or spectacle wearers in the target group). Whereas this is not a major problem for a large-scale scheme such as a passport programme, for a campus card it may be difficult to achieve reproducible conditions, particularly across a number of sites.

For nearly all biometrics, there is a small group of people who are unable to enrol successfully, and an alternative form of identification must be provided in these cases.

18.6.3   Registration, certification and verification

For any multi-application scheme, it is worth identifying which organisations and systems will perform the rôles of registration authority (RA) and certification authority (CA), and how user systems will relate to these.

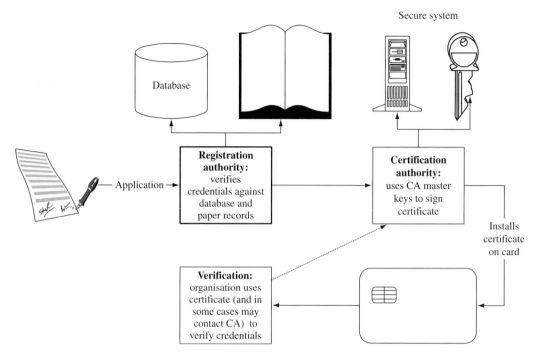

Figure 18.1 Registration, certification and verification

These terms are taken from cryptography (mainly public-key cryptography). The registration authority maintains the master database of certifiable information; it ensures that individuals only appear once, and that the data stored about those individuals are correct. If a biometric is used, the registration authority captures the biometric data and stores the template or a 'digest' of the template, to ensure that it cannot be altered. Most importantly, the RA *warrants* to all user systems that the data are correct – this may have liability implications where several organisations are involved.

User systems check the data by verifying the certificate issued by the certification authority. The CA holds the master (usually private) keys of the issuing organisation and uses these to sign the data passed to it by the registration authority. The CA function is mostly technical, and can be carried out using an in-house system or by a bureau, but the CA is also responsible for data security and key management. This process is illustrated in Figure 18.1.

18.6.4 Terminal management

Terminals in a campus system may be of various types: door entry systems, PCs on a corporate network, devices attached to photocopiers, vending machines, or payment terminals in a canteen or bar. Not all of these need be online: transactions may be collected periodically, by various methods, and some terminals may not need to gather data.

However, they must all be managed, and to perform a useful rôle in an authentication scheme they must contain some keys. Organisations often underestimate the effort required to manage an estate of terminals; a terminal management system (TMS) of some kind is generally essential. A TMS should include functions for identification of terminals and terminal software, key management, management of polling or other transaction collection, software and parameter downloading where appropriate, and the ability to disable any terminals that are abused or stolen.

18.6.5 Transaction processing

There is growing commonality around the use of IP (internet protocol) structures and XML message structures within transaction data collection, switching and processing schemes, so that very few systems have to be designed from scratch. In a campus scheme, there may be few constraints from external network or messaging standards, but it is worth bearing in mind any external networks to which the system may ultimately connect. To ensure that the system is flexible enough to allow organisations and departments to share the network, and to permit future expansion to new functions, message structures and data elements should be defined to reflect the rôles in the system as well as the immediate technical requirements.

18.7   Card requirements

As with any card scheme, the choice of card type represents a balance between cost, function and flexibility.

The first question is always: how much data must be stored on the card? If all transactions take place online, then most data can be stored in the central system. The card need only have enough data memory to support any offline functions, as well as the authentication-related information (templates and passwords should be stored in secret areas on the card, inaccessible through the Read record command). However, campus cards are notoriously prone to 'function creep' – having applications added to them – and it is greatly desirable to have enough memory in the card to store at least one extra application and its parameters; further extension beyond this point would be handled by an upgrade.

The physical features of the card are also important; many cards are personalised in-house using desktop machines, which work better with some card stock than others, and may also have some limitations as to the type of cards they can logically encode.

Campus cards are more likely to incorporate a multi-application operating system than many other card types, not only because the number of cards is likely to be smaller, but mainly because the applications themselves are much more often tailor-made and likely to change during the life of the card. The fact that multi-application operating systems generally support high-level languages and can download applications to the card after issuance makes them more attractive in this environment, despite their slightly higher unit cost.

18.8   The future of multi-application campus cards

Campus cards represent a balance between integration and distributed functions; it is likely that many of the card functions will at least be supported by functions and databases on a host system; for many applications the card performs a relatively small rôle and merely gives access to a host application.

However, in a world where organisations and users are increasingly conscious of security, privacy and the separation of domains, smart cards provide a uniquely strong mechanism for ensuring both security and privacy of information; their use in campus systems is almost certain to expand and to include a wider range of applications, including many that will require updating over the life of the card. The entry point for full smart-card management functions could fairly rapidly fall to the point where it is attractive to many large and multi-campus organisations, but the take-up from mainstream small and medium-sized businesses will depend on the availability of packaged solutions and turnkey software suppliers adopting this technology.

Part IV

Implementation

19 Organisation and structure

The previous five chapters have identified almost 100 suitable applications for smart cards, each with demonstrable benefits in some situations. Few people would imagine that all these applications could all reside on one card, although this would be technically possible, even today.

It is, therefore, useful to consider which applications have a good 'fit', so that they could share a card. Two perspectives are equally important:

- The card and application issuers must find it both easy and profitable to work together;
- For the card-holder, there should be a logical connection between the applications.

This chapter looks at each of these in turn: what makes it easy for organisations to work together and what are the barriers? And what 'domains of use' make sense to the card-holder?

19.1 Corporate culture

Probably the first factor that affects any co-operation or partnership between organisations concerns their corporate cultures and the personalities of the individuals involved. A large, slow-moving, risk-averse organisation is unlikely to make a good partner for a small, entrepreneurial business.

This, and indeed most of the organisational and commercial effects described in this chapter, can apply to multi-application projects involving several departments in one company, as well as to multiple companies. A successful co-operation in one country may not be transferable to other national operations of the same company.

In the case of card projects, attitudes to card-holders and users can vary widely, even within a sector. Sometimes the service offered to card-holders forms part of the service ethos in the company; in other cases the purpose of the card is to reduce customer servicing costs. The nature of the application can change the emphasis here: a customer-identification application is likely to belong to the IT or security department, while a loyalty or customer relationship application may be serviced by a sales or marketing department. These departments, in turn, may be more or less strongly influenced by the corporate culture.

19.2   Identifying stakeholders

In setting up any project, managers must have an eye to all those who are or may be affected by the project: this includes not only its direct users and customers, but also suppliers, special interest groups, trade associations, the press and government departments. It is valuable to have a strategy for handling each of these groups and addressing any issues or concerns they may have.

There is probably a good match if the stakeholders are similar for all applications on the card, and if the messages and proposition for each application to each group are similar. For example, it is good if all the applications stress the convenience and speed of using the card, but there might be a conflict if such a 'convenience' card also carries a secure ID application – it would be better to separate the applications in that case. A card primarily issued to disabled public-transport users could also carry a function allowing use of separate entrances or facilities in government buildings; this might be better than including that function on a card that offered access to a wide range of local and central government services.

Sometimes the presence of one application may complement, support or defuse an aspect of another: for instance, a card that incorporates the European standard for Coding of User Requirements (EN 1332–4) may provide value to the elderly or to disabled groups and thereby gain support from these sectors for the use of the card.

19.3   Trust hierarchies

Each sector has its own hierarchies. In the case of telecommunications, there are somewhat different structures for fixed-line services, mobile telephony and broadcast communications, but telecoms companies know where they stand in these structures and know how they can work with other operators. In banking, Visa and MasterCard run separate and competing hierarchies, which are separate from those run by SWIFT for inter-bank clearing. In the transportation business, most structures are at the national or even regional or city level, with few established contractual or legal structures that would support, for example, a public-key authentication system; such structures must be developed to meet the demands of a card scheme.

In the case of governments, there are strict hierarchies but these are sometimes changed by a new minister or by yet another local government reorganisation. For this reason, most governments set up a specialist agency responsible for authentication, independent of the underlying hierarchy. As a general rule, governments themselves do not admit of a 'higher authority' and so relationships between states must nearly always be on a peer-to-peer basis.

Many card schemes depend on an understanding of these hierarchies, and authentication functions often specifically depend on the existence of a common trusted third party, able to vouch for each of the parties in the system. In the absence of such

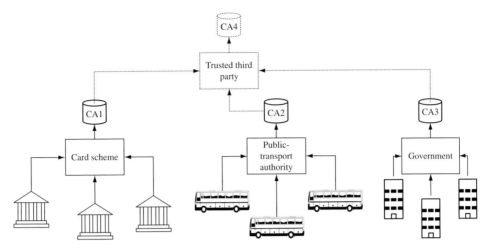

Figure 19.1 Trust structures with and without common hierarchy

a structure, contractual links and terms must be agreed between the parties, and a certification structure must be set up – see Figure 19.1. If this is external, there is an extra cost involved; if the card issuer acts as the certification authority (see Chapter 18) then all the other parties must trust the issuer.

One of the first obstacles to any multi-application scheme is the need for each party to understand not only the other parties' business needs, but also the environment and trust structures in which they work. If a common structure cannot be identified, then it will be difficult or impossible to set up a consistent scheme.

19.4   Liability

Wherever a service is provided, there are some potential liabilities. These may be trivially small, or may be made small by the terms of issue of the card, but even the smallest financial liability may have an impact on the image of a brand or on the trust that users place in it.

Most organisations are very risk-averse when it comes to the possibility of others damaging their brand or bringing a financial liability on them. They therefore seek to limit the potential damage and to describe and define all possible risks. In the very open-ended and unexplored world of multi-application cards, it is impossible to define all risks, and still less to quantify all potential liabilities. Very often, therefore, we end up with contracts written to transfer all risks away from the strongest party in the negotiation, which makes the terms unattractive for the other parties.

The only way round this is to define categories of risk that each party accepts, and other categories where the liability must be shared. This is often easier to agree for a pilot, or where one party's input is very small, than for long-term, equal partnerships of the type that smart cards were supposed to enable.

19.5   Commercial requirements

While the business case for any card project as a whole must be positive, individual applications may be seen as profit centres, cost centres or simply as part of a larger package. While a formal profit-centre approach is probably not common, it is very helpful to be clear at the outset what benefits (financial or other) are expected from each application, and what costs are attributable to that application. This is essential where the applications are owned or operated by separate organisations.

19.5.1   Co-branding

When a card does carry applications from several organisations, it will usually carry each of their brands in some form. There is a convention in banking cards that brands associated with the issuer are on the front of the card (the same side as the contact plate for a contact card), while acceptance marks (that show where the card may be used) are carried on the back. Although this is a useful convention, it is not followed by other sectors, who are more likely to view the front of the card as the prime space and the back as secondary: a logo on the front might be worth several times as much per mm^2 as a logo on the back.

Contactless cards overcome this problem to some extent, since there is no intrinsic difference between the two sides, and if there is no need to incorporate a magnetic stripe this gives the designers still more freedom. However, as the card is removed from the wallet less often, there is less benefit from the visual impact.

As mentioned in Chapter 13, owners are protective of their brands and of the characteristics associated with those brands, including image and colours. The relative positions of two logos may appear to make a statement about the relationship between the organisations. Co-branding agreements are fraught with difficulties such as these.

19.5.2   Rôles of partners

The other key factor in a multi-branded card is the rôles of the partners. In Chapter 9, I described the rôles defined by GlobalPlatform, and these are a good starting point, particularly for the technical functions associated with card issuance. There are also several commercial functions that must be allocated, depending on the nature of the scheme – for example, who provides the helpdesk for card-holders and for acceptors and users? Who is responsible for promoting the card to new card-holders, and for handling press enquiries? Who has responsibility for the security policies of the scheme as a whole? Are there any split rôles: areas where each partner handles some card-holders, for example?

These rôles and functions must be allocated at the outset, to avoid later problems. It is also very advantageous if the card-holder-facing rôles are clear to card-holders, and user or acceptor support rôles are clear to users – this avoids confusion and unnecessary cost or duplication.

19.5.3   Understanding each other's business

For these business needs to be met, and for rôles to be clearly defined, each party must understand the other's business. It must understand their motivation and priorities, the obstacles they face within their own organisations and externally, the rates of return and cost base they are using, and the operational processes that are linked to, or served by, the card.

This is very difficult: partners may not be prepared to disclose all this information, particularly at a negotiation stage. But it is dangerous to make too many assumptions: many people wrongly assume that they know how an organisation works just because they have contact with it as a customer. The sooner the parties can develop this common understanding, the better for the structure and organisation of the programme.

This can even be true within a single organisation: one department does not always know how another department works. Although, in this case, there should be no commercial barriers to developing a shared understanding, restrictive practices and old-fashioned attitudes may have an equally negative effect.

19.6   Card-holder 'domains of use'

In determining which applications have a good 'fit' together on a card, the card-holder's perspective is also important. Most people associate each card in their wallet or purse with one main function or group of functions, and an application that falls outside that group must be very compelling or it may well go unused. A card issued by an employer is a 'work' card, a public-transport card is associated with buses and trains – even if one of these carries a public telephone feature, users are more likely to buy a card from a telephone operator (unless the first card carries a good discount).

Operationally, we use different cards in different ways: some are kept as handy as possible, for regular use. Others are kept in a secure pocket, or not carried except when needed. Contactless cards work differently from contact cards, and often perform more reliably when carried separately from other cards, in an outside pocket of a jacket or handbag.

Multi-application card programmes work best when the applications actively complement one another (rather than simply sharing the cost of the card). Some examples of such domains of use include:

- Many people use one card for their whole journey to work, so it is useful to have functions on that card that help when making purchases on the journey: a newspaper or a cup of coffee. Some users would like to construct a 'journey card' that would include the parking, train, bus ride and newspaper, but not necessarily ad-hoc journeys, even on the same public-transport system.
- Shopping: most users understand that retailing is a competitive business and so will not expect all their shopping needs to be handled on a single card. A large minority of shoppers is happy to spend time exploring the options and finding out what extra

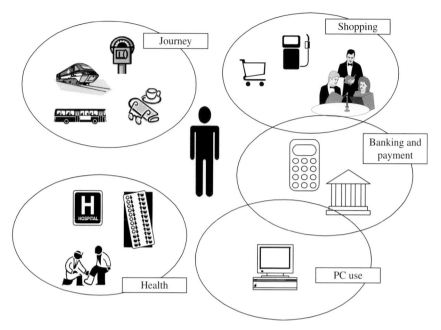

Figure 19.2 Card-holder domains of use

facilities are available – shopping comparisons are a serious activity for this group, and loyalty schemes an important part of this exercise. They will use a wide range of channels to claim rewards, discounts and special offers. An application that stores a record of previous purchases, including clothes sizes, colour numbers, etc., is a significant convenience and selling point for the card. It should also be possible to use the card for purchasing online from the same group of stores.

- Most PC users usually want to keep their card in the reader while they are using the PC, and then remove it when they leave or turn off. A card for PC use should, therefore, store not only login details and a user-authentication application, but also any other information that may be needed during the session: URLs and passwords for other sites, a digital signature application for e-commerce use, and user settings for common applications.

- Many customers would prefer their authentication tokens for e-banking and e-commerce to be stored on this same card (so that the card need not be removed to insert the bank-issued token or payment card) but this does conflict with banks' (and many customers') views of the security requirements for such applications. Such conflicts must be resolved before either set of applications is widely rolled out.

As Figure 19.2 shows, most applications fit reasonably well into one domain of use; however, there are some applications that can belong in two or more domains.

20 Implementation

In previous chapters, I have listed many potential barriers to a successful multi-application chip-card scheme, but also several 'enablers' – factors that will contribute to making a project successful.

This chapter is set out as a checklist; it sets these barriers and enablers against the phases of a project and should help those planning or implementing a multi-application project to make the right decisions at the right time.

20.1   Defining the project scope and road-map

For any project, clear definitions of scope and objectives are critical for success. In the case of a multi-application-card project, the *business objectives* must be clear and should drive the scope of the project. For the initiating organisation, its own business objectives must come first and should not be derailed by those of other partners.

There is a strong tendency for the scope of multi-application projects to grow during the life of the project: what is known as '*function creep*'. Whilst it is important to keep an eye open for changing circumstances, and some additional functions or markets may make good sense, any change to the original scope must be weighed against its effect on the business objectives and timescales.

Many multi-application projects will start with some basic features and add others as time goes on. It is very helpful to design a *road-map* that answers the questions:
– What do we have now?
– What will we add at each stage and how long will that take?
– What will we have at the end of this stage?
This reinforces the view that each additional function or group of functions must have a business benefit.

The functions and applications on the card must have a *logical fit*, from the card-holder's point of view. The card-holder must be able to put a mental label on the card: this is the card that I use for this purpose. Technologists may be tempted to put more functions on the card just because it is possible, or in the hope of saving card costs. For this to be successful, the card-holder's mentality and associations must be studied very carefully. Also project owners should beware of believing that they, and their way of thinking, would be representative of other card-holders.

20.2   Business case and risk analysis

There is an imbalance within many smart-card projects between those functions that are infrastructural (for example, securing systems) and those that are designed directly to yield revenue or to reduce costs. In drawing up the *business case*, it seems that many organisations work too hard on the direct cost-benefits and do not give sufficient weight to the infrastructure benefits. In some applications (for example, in banking) it is best to create two separate business cases: one for the necessary infrastructure changes, which do not seek to generate the threshold internal rate of return (IRR) required by many companies, and another for the revenue-generating projects. The first project is more likely to be justified by a risk analysis than by direct cost-benefits.

As we mentioned in Chapter 19, the *risk analysis* for a multi-application-card project must be undertaken from the top down, and must involve all partners and interested parties.

20.3   Choosing partners

Possibly the most tricky aspect of any project that involves multiple organisations is the *compatibility* between those organisations: there can be differences between departments within a single company, or between competing companies of a single type within a sector, but compatibility is most likely to be a major problem when the partners are in different sectors, have different corporate mentalities, or have completely different positions in their respective markets (for example, it is difficult for a small, aggressive company in a niche technology business to work with a traditional national monopoly – but many have tried).

It is important to ensure that both or all parties are working to the same *timescale* and are prepared to give the project the same *priority*, if need be (it may be easier for one than the other). *Decision-making processes* need to be aligned so that decisions, on operational matters particularly, can be made efficiently but without denying one party its say on matters that concern it. It is often useful to set up some 'what if ...' scenarios to test the compatibility of timescales, priorities and decision-making processes, and to document the outcome.

The choice of partners should *not limit the market* more than is necessary: sometimes working with one partner will limit the market to that partner's customers, with little prospect of extending it – it may mean all card-holders having to set up an account with the partner. Some sectors have specific issues here – for example, the need for banks to complete 'know your customer' checks for every new customer may delay account opening and hence card issuance. In the telecoms sector, working with a handset manufacturer to develop a product makes it very difficult to set up an arrangement with a network operator, and vice versa.

The partners need to be clear at an early stage what *rôles* they will perform, and as we saw in Chapter 19, those rôles that are customer-facing need to be clear to the customer as well. The rôles must not only be listed but also defined.

One of the advantages of a multi-application card with post-issuance downloading capability is that the issuer can consider *adding or changing partners* during the life of the card – a new application can be added or the parameters changed. However, if the old partner's branding was too prominent (on the card or other materials) or if it played too complex a rôle, then this may not be possible.

20.4   Identifying and managing stakeholders

It is important to identify all the *stakeholder groups* who will be affected by the project. Many projects suffer from identifying too few stakeholder groups, or from not segmenting those groups that they do identify. It is easy to see that card-holders will be stakeholders, but what about those whom the card is seeking to exclude, or who are excluded from one or more of the applications on the card? Front-end users, who will handle the card directly, are obviously stakeholders, but what about those who have to process errors or handle call-centre enquiries? How many suppliers will be involved? Are any of their interests threatened? Within each of these groups, it is likely that some members will be enthusiastic, some indifferent and some actively opposed. There needs to be a plan for identifying and managing each of these groups and segments.

Terminology can vary surprisingly widely from sector to sector, and sometimes even from region to region. The same person may be a customer of a telco, a card-holder or account-holder of a bank, a guest in a hotel, a passenger in a bus or train; depending on the legal niceties and politically preferred language, he or she may also be a member, citizen, subject or taxpayer, or all of these. It is important that the structure allows each application to use appropriate terminology.

20.5   Project organisation

The project *organisation* should reflect the rôles assumed by the different parties; once contracts have been let this may include suppliers as well as business owners, IT representatives and project managers. For most multi-application-card projects, marketing communications will be a significant enabling factor, and so some communications and marketing professionals are likely to be needed as part of the team from the start.

20.6   Timescales

Experience shows that the time required for several phases of a multi-application-card project is often underestimated. For example:

- Sufficient time must be allowed not only for approval by a board or higher authority, but also for one or more resubmissions giving more detail;

- Negotiation with partners is likely to be lengthy and it is difficult for the project initiator to impose a fixed timescale;
- Interfaces with existing systems must be defined in some detail, and the co-operation of incumbent suppliers may often be needed to deliver part of the system;
- There will be many phases of testing and certification, some of which the system will fail one or more times. Where the project initiator or lead issuer is not part of the 'club' or sector owning that particular certification or type-approval procedure, it will have limited ability to influence or speed up the process.

There is often a need for a balance between a simple procurement process, leading in most cases to shorter implementation times, and the degree of control the project owner can exercise over the final outcome. Many complex card operations are out-sourced, in an effort to minimise unnecessary delays.

20.7   Standards and specifications

Most sectors are able to refer to a standard for their core application, as described in Chapters 14–17. However, there is often no standard for any secondary application (such as loyalty) and for campus cards the only standards relate to the host applications, not the cards. Most of the standards that do exist permit quite a wide range of options, and not all vendors' products will support all options.

So there is always a need for a *detailed specification*, even if that specification refers frequently to the standard or standards. In principle, each application owner is responsible for its own application specification, but in practice most multi-application-card projects demand a moderately high level of information sharing between card issuer and application owner; there may be some parameters that remain firmly under the application owner's control, but within limits understood by all parties.

In years gone by, the problem with smart cards was a lack of standards. This could not be said today, although there are still many applications where further standardisation would be beneficial. One new problem is the *rate of change* of the standards that do exist: many suppliers cannot keep up with all the changes and the certification involved, with the effect that some applications refer to old versions of the standard. Multi-application cards at least allow the possibility of upgrading to a newer version, provided there are no hardware obstacles.

20.8   Procurement

Purchasing smart cards is a specialist business that appears to break many of the rules of procurement. The lowest-price product is not necessarily the one with the fewest functions, nor even does a higher volume always secure a lower price. The lowest-price product is usually that which has recently been in highest demand, and so has been allocated to the highest-volume production lines – provided, of course, that demand does not exceed supply, in which case the price can rise sharply in a short time.

Smart-card manufacturers prefer orders that secure production over an extended period – they reward forward orders and delivery schedules that do not change. From the purchaser's point of view, however, the need to ensure continuity and security of supply (there are sometimes shortages of particular chip types) must be balanced against the desire for flexibility in product design and marketing.

Using a multi-application card gives significantly more flexibility and allows quite a wide range of products to be delivered using a single card type. But the difference in price between a native and a multi-application card can be such that it is cheaper to use a native card and re-issue it if further applications are required, while the continuous upgrading of specifications and new features offered by multi-application operating systems mean that cards do become out of date. This is particularly important for applications such as mobile telephone SIMs, where the demand for memory is increasing rapidly.

20.9   Operational process design

It is as easy to design bad processes for using smart cards as it is to design good ones. Simple things like the orientation of the reader, the colour and size of instructions and shaped slots can not only make life much easier for unfamiliar or disabled users, but also often speed up transaction times and improve reliability for everyone.

For a large-scale scheme, both market research and ergonomic design techniques should be used in an effort to find the best physical and psychological presentation.

20.10   Managing risks, problem and learning

During the project there must be a continuous process for identifying risks and problems, grading them as to their seriousness and likelihood, and finding a route to their resolution. This will often involve sharing information between participants, and co-operating in the issue resolution.

There is always a danger with a multi-application project that a solution to one problem also impacts other areas; sharing the problem and risk register not only helps to avoid nasty surprises but also allows each partner to learn from the experience of others.

Case study P – UK Chip and PIN programme

Chip and PIN

Organisation

The UK Chip and PIN programme was set up by the Association for Payment Clearing Services (representing the banks that issue cards and acquire transactions) and the British Retail Consortium (representing the retail industry). It was a unique co-operation between two sectors that have often been at loggerheads in the past, and the first task of the team was to design a committee and governance structure that would allow decisions to be made following suitable consultation.

The Programme Management Office (PMO) was a co-ordinating team only, with no direct control over any of the individual projects of the banks or retailers. However, it was able to demonstrate its independence and authority and this greatly contributed to the success of the programme.

Objectives

At the start of the programme in January 2002, there were 30 million chip cards and 90 million magnetic stripe cards issued by the banking industry, but all of these used signature verification in shops; at ATMs the magnetic stripe was used, together with a PIN, which was checked online by the host system. The objective of the programme was to pilot and then introduce chip and offline PIN technology[1] across the UK, in such a way that an irreversible critical mass had been reached by 1 January 2005, when the international card schemes would introduce a change in the liability rules for fraudulent transactions. This was the first national roll-out of the new international EMV standards.

Structure

The Programme Management Office was structured into three workstreams:

Technology and operations

Although the standards for chip and PIN cards were set by the international card schemes, there was a need to define more precisely the options that would be followed in the UK and the way in which they would be implemented, to provide some consistency for retailers and card-holders. These took the form of 36 recommendations and guidelines, which were subsequently incorporated into APACS Standards.

There were also many issues and questions raised about interoperability: with 30 or 40 suppliers of bank cards and systems, and many hundreds of suppliers to the retail industry, it was not possible to test every combination, so a mechanism for resolving problems as they arose had to be developed.

Communications

The success of the programme depended on card-holders (consumers) and retail staff changing their behaviour. Since many card-holders have cards from more than one bank, and they all shop in several outlets, consistency of messages was important. This was achieved by a combination of agreed messages delivered by each participant (bank or retailer) to its customers, and central communications delivered through television, radio and print advertising and a co-ordinated PR campaign.

---

[1] The PIN being checked on the card.

The communications team was also responsible for stakeholder management: identification of interest groups, market research and targeted communications, where required.

Implementation
Delivery within the required timescales required participants to work in parallel and, therefore, to have confidence that others would achieve their timescales. Although the PMO was not responsible for managing individual participants' projects, it monitored all those projects and identified key milestones by which certain phases must be complete.

Consultative fora
Alongside the workstream committees, consultation groups were introduced for vendors, disability groups, trade associations and others with an indirect interest in the programme.

Technology
The technology adopted by nearly all card issuers was a single-application, native-operating-system card with a small number of datasets. This minimised the implementation risk and card cost, while any further applications could be added when the cards were renewed. The terminals installed by banks generally had a multi-application capability, but few utilised anything other than the basic payment application. However, those retailers who owned their own payment equipment more often chose to integrate the payment software into their point-of-sale terminals, alongside other applications, and this meant that each retailer's system had to be certified by the card schemes and acquiring bank.

Challenges
Among the many problems addressed by the PMO, the management of disability groups was very important. Although many disabled card-holders benefited from the transition, this group needed to be assured that its interests were being taken into account as far as was possible.

The PMO sometimes needed more information to manage the transition than the participants had specified to manage their ongoing business. There was a need to develop management information systems specifically for the transition and introduction phase.

In response to the shortage of test cards (see main text), a combination of exhaustive testing using test-card packs, with a special test room where participants could test live cards on live terminals, gave sufficient confidence and minimised interoperability problems.

Outcome
A pilot was run in one town starting in May 2003, mainly to test the consumer-facing operations and messages, and the main roll-out started in October 2003.

By the end of 2004 approximately 70% of cards and 65% of terminals had been converted to chip and PIN, and by February 2006 it was possible to remove the option for UK card-holders to use signatures with chip and PIN cards.

The next stage for many banks was to use the cards for authenticating e-banking and e-commerce customers; a standard for this was delivered in late 2005 and the first implementations were launched in mid 2006.

20.11   Testing

One frequent problem with smart-card projects is a shortage of test cards. Since the card changes each time it is used, tests with real cards are not repeatable; but tests with simulators may not accurately represent real-life behaviour. Card and terminal applications, personalisation links and host systems are often developed in parallel, which means that there are no cards available for unit and function testing of the terminals and host systems.

There is no complete solution to this, but carefully designed test cards and simulators do help considerably, and project owners will usually find that developing or buying in such simulators reduces the overall project timescale considerably.

20.12   Going live

Clear, bold literature is likely to be needed at all points of use, in the period leading up to the introduction ('This is what we are going to do and why we are doing it.') as well as during the introductory period ('This is what you should do now.').

Nonetheless, in the early stages of introducing or migrating to a multi-application smart card, there is always some confusion. New stakeholders are identified and new questions asked – if a ready answer has not been prepared then staff will invent an answer, which may not be correct. A technique that has been found to help considerably is to issue the new cards first, wherever possible, to all staff who will have to answer these questions – these include users such as retail and security staff, bus and platform staff, as well as call centre and helpdesk operators.

20.13   Communication

Throughout this process, regular or appropriate communication with all stakeholder groups is needed (for a new product launch it is generally not possible to start advising customers ahead of the launch). For a major project, such as a national ID card, it is worth testing all messages in advance to ensure that they will have the right impact when they are released.

Countering negative reactions is also important and requires a measured approach: a small number of negative voices can seem disproportionately loud, particularly to a project owner who is proud of his or her 'baby'.

21 Prognosis

21.1 Technology

It will be clear from the rest of this book that the availability of technology is no longer a limiting factor preventing the deployment of multi-application smart-card schemes. However, further technology developments will continue to appear and some of these will be distinctly helpful by allowing a wider range of applications, or by making existing applications work better or at lower cost.

21.1.1 Microcontrollers

At the chip level, semiconductor technology as a whole continues to advance in line with Moore's Law: doubling the number of gates per chip every eighteen months. In the case of microcontroller chips, the 0.12–0.15 µm technologies that are regarded as leading-edge in 2006 are believed to be close to the limit for E^2PROM; however, flash memory is being used to grow total memory sizes into the megabyte range, and this technology will be used increasingly in combination with E^2PROM to provide the memory sizes required by the telecommunications industry today, and probably for multi-application cards in the near future.

In 2000, I forecast that smart-card microcontrollers would be using 0.1 µm processes by 2005; this turns out to have been optimistic, but this level is now regularly used for DRAM products and should be achievable by 2007 for microcontrollers, moving to 0.07 µm or less by 2010.

Memory sizes for microprocessor cards, currently mostly in the range 4–128 kB, are likely to rise over the next few years to 32 kB–8 MB, with a wider range of combinations of memory types, perhaps configurable at a relatively late stage in the manufacturing process.

Various forms of co-processor will be introduced, not only for advanced mathematics and cryptographic applications, but also to assist other security, organisational and communications tasks. There will be more specialised language-set processors, including chips that implement Java and MEL directly, as well as conventional RISC machines optimised for performance. Some chips are already tailored to either

JavaCard or Multos, and we must expect further optimisation and specialisation in this area.

21.1.2   Cards

Card technology is well established and stable, but must still keep running to stay ahead of fraudsters and other pressures. Many card types (particularly bank cards) have hitherto been limited by the need for backwards compatibility, going right back to embossing in many cases. As more transactions become electronic, this need for embossing, and even for magnetic stripes, will diminish. Most other applications are relatively free of such legacy issues, and can use the best currently available technology.

There are already several materials competing with PVC for use in cards. Multi-application cards must often be very durable, with a life of up to ten years; PETG and polycarbonate are often used as carriers for these card types, although they are difficult to emboss.

Embossing is also unfriendly to contactless cards, which need the whole circumference of the card for best performance of the antenna; smaller loops do work but couple less power at any given distance. For multi-application cards, contactless interfaces will become much more common and they may even become the norm for stand-alone cards by the end of the decade.

Contactless cards have many fewer limitations as to form factor, so we will also see many new card shapes, appropriate to the type of application: belt clips or buttons for travel, pens for work, battery covers for mobile phones, necklaces or bracelets for ID, etc. The one sector that will struggle (for reasons of certification and backwards compatibility) to work with contactless cards and non-standard form factors is banking; this could lead to a divergence between bank cards (which may still be multi-application but limited to that sector) and other sectors which can more readily share a multi-application contactless 'card'.

21.1.3   Terminals

Now that the card-to-terminal protocols have been more or less standardised, more terminals will start to include contactless interfaces as well as contact; the chips to process the $T = CL$ protocols as well as $T = 0$ and $T = 1$ are already available, and the higher-level message handling can be done in software. More contactless-only terminals will be developed; being free from form factor constraints, they can be designed to fit vehicles, arm-rests, fast-flow gates, turnstiles, and even for use at points of sale without occupying counter space (always a problem for conventional payment terminals).

There will be a growing number of 'personal terminals', ranging from mobile telephones with NFC interfaces to smart phones and portable financial terminals. Whereas we are forecasting that cards and mainstream terminals will increasingly be multi-application, many of these personal terminals may be dedicated to a particular service.

21.1.4 Card and terminal management systems

Card and terminal management systems will become increasingly important; however, instead of the current trend towards ever-more-powerful systems, with a very wide range of features appropriate to all industries, there will need to be a move towards functional modularity (the ability to select which functions are required for each scheme), to avoid these systems becoming unmanageably large and over-burdened with certifications.

There may be two divergent trends: one towards single-sector products optimised for, and highly responsive to, the needs of that sector, and another towards more general products that incorporate interfaces or hooks for sector-specific functions. It is difficult to predict which approach would be more successful.

21.1.5 Security

Smart-card security continues to advance steadily, while at the same time the range of attacks available to the determined attacker grows continuously wider. However, the cost of mounting these attacks also rises, while the benefits of penetrating a single card are limited by the wider system of which the card forms part.

We will see further advances in chip and card security, triggered partly by the card companies' own research and partly in response to threats discovered by outside laboratories. However the most important response to the security threat will be closer co-operation and integration between the card and the wider systems of which it forms part – the systems must remain secure even if the card is penetrated.

21.1.6 Standards

It seems unlikely that the current profusion of standards covering all aspects of smart cards and their application will be rationalised – more probably, they will be joined by new standards, specifications and certification requirements.

Without reducing the number of standards, some order and easier implementation could be brought to the scene by introducing a consistent layering scheme into these standards, showing which are true alternatives, which cover different aspects of the technology, and which may simply co-exist to handle different sectors or functional requirements. This could be introduced by an organisation such as GlobalPlatform or StepNexus.

21.2 Applications

There is a growing number of applications using smart cards, and increasing awareness of the technology leads to more demands. The availability of contactless and 'combi' cards enables or helps several new applications, while for other applications it makes no difference.

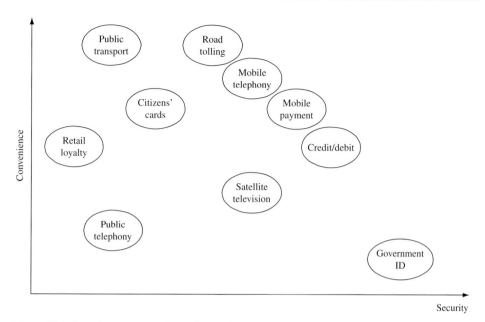

Figure 21.1 Security vs. convenience for typical applications

Smart cards will increasingly be used for verifying our identity, whether at borders, at retail points of sale or when shopping online or signing on to a PC system. Unconnected networks and incompatible trust structures will continue to prevent some schemes from growing and interoperating, but the effect of these limitations will probably be to increase the number of schemes rather than decrease it.

Linked to ID will be not only physical access but also database access: for local government, healthcare and a wide range of campus applications. Most forms of payment are actually a form of secure database access also, but with additional data provided by the card: the stored-value card or electronic purse seems to be less important than it was but is not to be written off yet.

Figure 21.1 addresses the main drivers for some typical applications, particularly the extent to which they are driven by the need for security or to offer convenience to the user; this is a helpful indicator of the natural affinity between applications but also shows why each chooses the technology it does.

21.2.1 Key sectors

Key sectors for developing or leading multi-application smart-card projects are those that have a strong business case or rationale for adopting the technology, or for whom there is no other solution to a pressing security problem. High on this list are governments and transport companies, the first for security reasons and the second for cost control and customer service. In many countries, large parts of the transport infrastructure are in fact government-owned, and so the two sectors may be linked.

Telecommunications and banking are two other sectors that have a strong motivation to use smart cards, and there is some convergence between these two applications. They are also the applications that are more likely to use contact cards, and so banks and telcos may find it easier to share a card than either of these with a transport card or government department.

21.2.2 Inter-sector and intra-sector cards

Despite all the systems available to facilitate inter-sector working, it remains easier for organisations within a broad sector to work together than to work with an organisation from another sector. In this book, I have mostly considered four broad sectors: telecommunications, banking and finance, transportation and government. In addition, the classification 'campus cards' covers any organisation in its rôle as employer, as well as educational establishments.

We have seen that it is often difficult for organisations from different sectors to share a card, for reasons which range from the different hierarchies within which they work, through cultural, marketing and branding issues, to the need for certification of cards, terminals and applications by the relevant authorities within the sector. Intra-sector sharing, by contrast, is subject to one issue: competition.

The best candidates for cards shared between organisations are, therefore, where the sector is large enough and competition is weak or non-existent. This explains the predominance of governments and government-owned bodies in the case studies in this book. Within the transportation business, there are also many non-competing providers: the public-transport authorities in different cities do not compete with each other, with the national airline nor, in most cases, with a long-distance train operator. So this is another sector where multi-application cards may carry applications from many service providers.

In most countries, however, telecommunications and banking are highly competitive sectors, where branding and customer relations are critically important, but where the range of services offered is also a key differentiator. It is, therefore, very likely that multi-application cards in these sectors will be single-issuer but function-rich. Figure 21.2 shows some possible scenarios for the evolution of applications and multi-application card types over the coming five to ten years.

Historically several technologies and approaches to implementing a multi-application scheme have been closely associated with one sector: campus card schemes have used host-based systems; Multos cards are most popular in government applications; the telecoms industry favours JavaCard technology and is the only one to use dynamic updating of applications on any scale; banks issue native-operating-system cards but have relatively high demands for security certification, while transport cards have been slow to migrate from wired-logic to microprocessors but very quick to adopt contactless technology.

The growth in each one of these sectors clearly shows that all the technologies described work and all meet some aspect of users' requirements. To benefit from multi-application schemes and, particularly, schemes that cross sector boundaries,

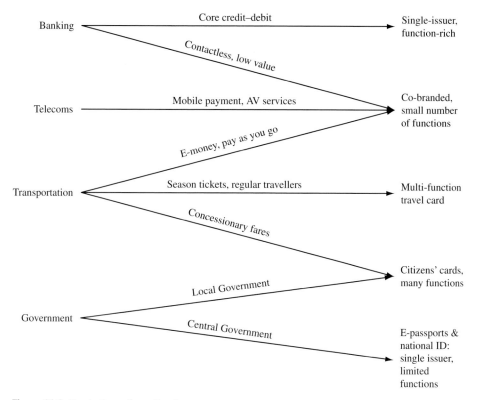

Figure 21.2 Evolution of applications

operators must recognise the value that each of these different approaches brings to its users. All sectors must, therefore, look more closely at other smart-card users and learn from the experiences of others. They must shop around to select the most appropriate standards and specifications, and mix technologies and techniques to create an effective system. For example, bankers seeking to deploy contactless cards can learn much from public-transport operators, while governments looking to deploy effective communications channels towards their citizens could look to telcos for a model.

21.3   Towards a more customer-focused view

Multi-application cards provide a rich seam both for technologists, who like to see technology exploited to its maximum (tiny geometry, complex mathematics, highly distributed networking), and for marketing people, who can use their imagination freely to invent new ways to use the card and exploit its marketing potential. Too often, both these ideals come to grief when confronted with the financial controller, who seeks a solid business case.

A more constructive approach for both groups would often be to explore the customer relationship and discover the combination of core functions, additional services, brand image, trust and controls that embodies this particular relationship (and which may vary from customer to customer). The card can then be viewed as a way of facilitating or enhancing that relationship.

This helps to discover whether the card must be dedicated to the issuer's functions or whether applications from other issuers have a place; and if so whether they should sit alongside or under the card issuer's applications. It helps to work out the logical grouping of applications between cards and within cards, and how cards should be positioned and marketed to the customer.

Even for governments and others who have a captive user base, defining the card in terms of what it adds to the organisation's relationship with the card-holder helps to define the functions of the card in a way that will subsequently ease its introduction.

In considering some of the more complex schemes that have been proposed recently, issuers have had to address several of these questions, and there are signs that many of them are now starting to move in this direction. This should be healthy for the smart-card industry, for its hardware and software suppliers, integrators and promoters; but most of all it should be beneficial to scheme operators and to those who are planning to set up a multi-application or multi-function card scheme.

Appendix A

Glossary

AID	Application identifier: the registered name used for application selection.
Anti-collision	A protocol that allows a contactless reader to identify and transact correctly with each of several cards in the reader's field at once.
APDU	Application protocol data unit: the structure used for commands and responses exchanged between terminal and card under ISO 7816-4.
API	Application programming interface: a specification that sufficiently defines an interface (to a device, for example) for a programmer to develop code that will drive the device, without needing access to the device.
Application	In the smart-card world, the word 'application' is often used for the program on the card or terminal that performs a function or provides a service; in the case of a card, the application may be burnt into ROM (the 'hard mask' of the card) or stored in E^2PROM or other rewriteable memory.
ATM	In banking, an automated teller machine or cash machine.
ATR	Answer to reset: the data sent by a card in response to a reset signal.
bps	Bits per second.
CA	Certification authority: the system or organisation that electronically signs a package of data using a private key, to generate a certificate; see also RA. In satellite and cable television, conditional access: system to prevent unauthorised viewing of transmissions.
Calypso	European standard for interoperable transport ticketing.
CAP	In JavaCard, the converted applet file or cardlet package contains the classes in the format required for loading to the card.

	MasterCard also uses this acronym for its card-holder authentication program, used for authenticating customers using e-banking or e-commerce.
CDMA	Code division multiple access: a range of techniques for sharing bandwidth between several data streams, used in both second and third generation mobile telephony.
Cloning	Making an identical copy of a card; in practice this term is often used if the copy appears identical to the original to the outside world, perhaps only under some circumstances (such as offline transactions).
Digest	A digest includes the most important elements of a message (or other data) such as the message number, date and time, and main data fields; it is often *hashed* (q.v.) and used to construct an *MAC* (q.v.).
Digital signature	Authentication of data by the originator using a private key.
DPA	Differential power analysis: a technique for deriving information about keys and other card data from analysis of the power input of a smart-card chip.
DRAM	Dynamic random access memory: memory that can be accessed directly but must be powered continuously to retain data.
EAL	One of the seven evaluation assurance levels specified in the Common Criteria for Information Security.
E^2PROM (or EEPROM)	Electrically erasable read-only memory.
EMV	The international standard for smart credit and debit cards, owned and managed by EMVCo.
ETSI	European Telecommunications Standards Institute.
FINREAD	Specification for a secure personal smart-card reader device with keypad and display, published by the European Standardisation Centre (CEN) as CEN Workshop Agreement (CWA) 14174.
GSM	Global System for Mobile Communication: international standard for 'second-generation' digital mobile telephony.
Hash	A hash is a one-way function that reduces a block of data (of any size) to a predetermined length; the original data cannot be reconstructed from the hash value. Hash functions are often used as part of a message authentication scheme.
IATA	International Air Transport Association.

IFD	Interface device: an alternative term for a smart-card reader.
IP	Internet protocol: the device addressing protocol used on the internet and many other networks. The IP version 4 was limited to 32 bits (usually expressed as four decimal numbers, e.g., 255.255.255.255), but version 6 extends this to 128 bits, as well as removing many other limitations of IP V4.
IPR	Intellectual property rights.
ITSEC	Information Technology Security Evaluation and Certification scheme: the European scheme for evaluating and certifying the security of IT products, now being superseded by Common Criteria for IT Security Evaluation (ISO 15408).
ITSO	Originally the Integrated Transport Smartcard Organisation, ITSO now prefers to be called by its initials as it is not restricted to transport or to smart cards.
kbps	Kilobits per second.
KMA	Key Management Authority: generates pairs of public key, manages master keys and signs certificates for an organisation or function.
MAC	Message authentication check: a field on the end of a message that proves that it came from the person or unit identified in the 'from' field and has not been altered. When public-key encryption is used to generate the MAC, the result is usually known as a digital signature.
MI	Management information.
NFC	Near field communication.
NIST	National Institute of Standards and Technology.
Not-on-us	See On-us.
OEM	Original equipment manufacturer: OEM equipment is produced by one company but repackaged and sold by a second; often it is designed to be incorporated into a larger integrated system.
On-us	A transaction where the originating bank is the same as the destination bank: the card-holder uses the same bank as the shop, or draws money from his or her own bank's ATM.
PCD	Proximity coupling device: the term used by ISO 14443 for a contactless card reader.
Personalisation	Loading the data and any software applications into the memory of a smart card – often the process is divided into two stages: programs are loaded first and then the user-specific details such as name, account number or a biometric are added.

PIN	Personal identification number: a password of, usually, between four and eight numeric digits.
PP	Protection profile: a set of security requirements selected for a particular application and to be met by any implementation.
PPS	Protocol and parameter selection: the protocol by which a card and terminal agree a communications protocol and speed.
RA	Registration authority: the system or organisation that warrants the correctness of the data contained in a certificate; see also CA.
RF	Radio frequency.
RFID	Radio frequency identification: any of the technologies that allow a tag or other object to transmit an identifier or similar data to a reader using RF transmission. This term covers communication over distances of a few mm to several metres; in the USA, it is often used to refer to any contactless smart-card technology.
RWU	Read–write unit: an alternative term for a smart-card reader.
SIM	Subscriber identity module: the card that maintains the secure user authentication and further user data in a GSM telephone; see also USIM.
TCSEC	Trusted Computer System Evaluation Criteria: the US equivalent of ITSEC (q.v.).
Tearing	Where the card is removed from the reader before the transaction is complete.
URL	Uniform resource locator: the name of a resource (most often a web page or device) on the internet.
USIM	Universal subscriber identity module: the card that maintains the secure user authentication and further user data in a 'third generation' (3G) telephone.

Appendix B

Further reading

B.1 Smart-card technology

Everett, D. Smart Card Tutorial. *Smart Card News 1992–1994* (downloadable
at www.smartcard.co.uk/tutorials/sct-itsc.pdf)
Finkenzeller, K. *RFID Handbook*. 2nd edn, Wiley 2003
Hendry, M. *Smart Card Security and Applications*. 2nd edn, Artech House 2001
Rankl, W. and Effing, W. *Smart Card Handbook*. 2nd edn, Wiley 2000; 3rd edn, Wiley 2003,
also available as e-book

B.2 Biometrics

Bolle, R. *et al*. *Guide to Biometrics*. Springer-Verlag 2003

B.3 Cryptography and card security

Anderson, R. A range of papers on reliability of security systems is listed at www.cl.cam.
ac.uk/~rja14
Ferguson, N. and Schneier, B. *Practical Cryptography*. Wiley & Sons 2003
Kahn, D. *The Codebreakers: the Comprehensive History of Secret Communication from Ancient
Times to the Internet*. Simon & Schuster 1996
Menezes, A., van Oorschot, P. and Vanstone, S. *Handbook of Applied Cryptography*. 5th edn,
CRC Press 2001 (a very mathematical treatment)
Piper, F. and Murphy, S. *Cryptography: a Very Short Introduction*. Oxford Paperbacks 2002
Purser, S. *A Practical Guide to Managing Information Security*. Artech House 2004
Schneier, B. *Applied Cryptography*. 2nd edn, Wiley & Sons 1996
Schneier, B. *Secrets and Lies*. Wiley & Sons 2000

B.4   JavaCard

Baentsch, M. *et al.* JavaCard – from hype to reality. *IEEE Concurrency*, **7**(4), 36–43, 1999

Chen, Z. *JavaCard Technology for Smart Cards: Architecture and Programmer's Guide.* Addison-Wesley 2000

Hansmann, U. *et al. Smart Card Application Development Using Java.* 2nd edn, Springer-Verlag 2002

Appendix C

Standards

This appendix lists the most important international standards applicable to multi-application smart-card systems. As described in the text, there are many more standards applicable within individual sectors.

Standard name	Title and description
ANSI X3.92	*Data Encryption Algorithm (DEA)*. The main source for the DES algorithm.
ANSI X9.19	*Message Authentication Check*. The most widespread form of symmetric-encryption MAC.
ANSI X9.30-2 (1993)	*Part 2: Secure Hash Algorithm (SHA-1)*. Public-key cryptography using irreversible algorithms for the financial services industry.
ANSI X9.31-1	*Part 1: The RSA Signature Algorithm*. Public-key cryptography using reversible algorithms for the financial services industry.
	RSA is, by some margin, the most widely used public-key encryption algorithm. Although the algorithm is patented, RSA Data Security licenses it for general use and this standard gives recommendations in this respect.
CEPS	*Common Electronic Purse Specifications*: a set of business requirements, technical and commercial specifications covering interoperability of contact-based electronic-purse cards. This standard is issued and maintained by CEPSCo LLC: www.cepsco.org, but is now little-used, as electronic-purse activity moves to contactless cards.
EMV	*Integrated Circuit Card Specification for Payment Systems*. The international standard for communication between chip-based credit and debit cards and their terminals. The standard is managed by EMVCo: www.emvco.com

Standard name	Title and description
EN 726	*Identification Card Systems – Telecommunications Integrated Circuit(s) Cards and Terminals.* This standard was produced by the ETSI working group TE9; its seven parts cover all aspects of the design of a multi-function IC card with the exception of the application itself.
EN 1038	*Identification Card Systems – Telecommunications Applications. Integrated Circuit Cards for Payphones.* This standard, which was produced by CEN, complements EN 726.
EN 1332	*Identification Card Systems – Man-machine Interface* *Part 1: General Design Principles* *Part 2: Card Orientation* *Part 3: Keypads* *Part 4: Coding of User Requirements* Part 2 defines the 'notch' to help blind people orient a card correctly. Part 3 defines the layout (telephone-style), tactile identifier, colour and arrangement for keypads on card-operated devices. Part 4 defines how details of a user's preferred interface can be stored on the card. These preferences could include large characters on the screen, speech prompts, more time for key entry or amplification of sound output.
EN 1546	*Identification Card Systems – Inter-sector Electronic Purse.* A development from EN 726 and EN 1038, but with banking industry as well as telecommunications industry involvement.
ETSI	The European Telecommunications Standards Institute publishes the specifications for the GSM and 3G Partnership Project – see www.etsi.org/services_products/freestandard/home.htm and http://pda.etsi.org/pda/.
FIPS 201	*Personal Identity Verification (PIV) of Federal Employees and Contractors.* Although this is a US standard, it is having a growing impact on personal identification requirements worldwide.
IATA	The International Air Transport Association's Resolution 795 covers an electronic ticket to be held in a smart card.
ISO[1] 7810 (1985)	*Identification Cards – Physical Characteristics* This standard describes the shape, size and environmental requirements of a plastic card to be used as an identification card. In practice, this standard size of card (85×54 mm) is used in almost all card applications.

Standard name	Title and description
ISO 7811	*Identification Cards – Recording Technique.* The six parts of this standard cover embossing and magnetic stripe recording.
ISO/IEC 7816 (1998–2005)	*Identification Cards – Integrated Circuit Cards with Contacts.* The 15 parts of this specification cover all aspects of inter-sector smart cards, including shape and size, layout and use of the contacts, communications, memory organisation, application selection, security, etc.
ISO 8583 (1998, 2003)	*Financial Transaction Card Originated Messages – Interchange Message Specifications.* Although primarily intended as an inter-bank messaging specification, this standard is used as the basis for most terminal-to-host messaging for financial transactions.

ISO 8731	*Banking – Approved Algorithms for Message Authentication*
	    Part 1: Data Encryption Algorithm
	    Part 2: Message Authenticator Algorithm

ISO 9564 (1991)	*Banking – PIN Management and Security* The two parts of this standard define the techniques and approved algorithms recommended for handling PINs securely in a retail banking environment.

ISO 10202	*Financial Transaction Cards – Security Architecture of Financial Transaction Systems Using Integrated Circuit Cards*
	    Part 1: Card Life-Cycle
	    Part 2: Transaction Process
	    Part 3: Cryptographic Key Relationships
	    Part 4: Secure Application Modules
	    Part 5: Use of Algorithms
	    Part 6: Card-holder Verification
	    Part 7: Key Management
	    Part 8: General Principles and Overview
	Part 6 is mainly concerned with PIN verification, although an annex allows for passwords or biometric methods.

ISO 10536	*Identification Card – Contactless Integrated Circuit Cards:*
	    Close-coupled Cards
	    Part 1: Physical Characteristics
	    Part 2: Dimensions and Location of Coupling Areas
	    Part 3: Electronic Signals and Reset Procedures
	    Part 4: Answer to Reset and Transmission Protocols
	This standard defines a 'close-coupled' card (up to 10 cm).

ISO 11568 (2005)	*Banking – Key Management (Retail).* The five parts of this standard define the principles, techniques and life-cycle of keys in a financial card environment.

Standard name	Title and description
ISO 14443 (2000–2005)	*Identification Cards – Contactless Integrated Circuit(s) Cards – Proximity Cards* This specification covers the two most common contactless card types used today.
ISO 15693 (2000)	*Identification Cards – Contactless Integrated Circuit(s) Cards – Vicinity Cards* This standard covers cards for use at greater coupling distances than the other two contactless standards (up to 1.5 m).
ISO/IEC 18033-3	*Information Technology – Security Techniques – Encryption Algorithms*. Part 3 of this standard defines the use of the triple-DES algorithm.
ISO/IEC 18092	*Telecommunications and Information Exchange Between Systems – Near Field Communication – Interface and Protocol (NFCIP-1)*
ITSO	Specifications originating in the UK transportation sector, but which describe a fully interoperable scheme for sharing smart-card products (e.g., tickets) across operators in a distributed network, under a common security infrastructure.
MRTD	The standards set by the International Civil Aviation Organisation (ICAO) for machine-readable travel documents (e-passports and e-visas) are available at www.icao.int/mrtd/
PC/SC	The PC smart-card specifications are a group of standards for using smart cards in a PC environment – see Figure 7.3 in the text.

[1] All ISO standards can be obtained from the International Organization for Standardization at www.iso.org

Index